# INTERNAL COMBUSTION ENGINES

## A Reference Book for Designers, Operators, Engineers, and Students



MAXWELL PRESS

# INTERNAL COMBUSTION ENGINES

A Reference Book for Designers, Operators, Engineers, and Students

**William Manville Hogle B.S**

Consulting Engineer

MAXWELL PRESS

Chennai Trichy New Delhi

MAXWELL PRESS

This edition has been published in india by arrangement with Carson Books, UK

ISBN 978-93-5528-206-4

**Maxwell Press**
No. 44, Nallathambi Street,
Triplicane, Chennai 600 005



MJP 1282



Publisher : C. Janarthanan

# Publisher's Note

The legacy of a country is in its varied cultural heritage, historical literature, developments in the field of economy and science. The top nations in the world are competing in the field of science, economy and literature. This vast legacy has to be conserved and documented so that it can be bestowed to the future generation. The knowledge of this legacy is slowly getting perished in the present generation due to lack of documentation.

Keeping this in mind, the concern with retrospective acquiring of rare books has been accented recently by the burgeoning reprint industry. Maxwell Press is gratified to retrieve the rare collections with a view to bring back those books that were landmarks in their time.

In this effort, a series of rare books would be republished under the banner, "Maxwell Press". The books in the reprint series have been carefully selected for their contemporary usefulness as well as their historical importance within the intellectual. We reconstruct the book with slight enhancements made for better presentation, without affecting the contents of the original edition.

Most of the works selected for republishing covers a huge range of subjects, from history to anthropology. We believe this reprint edition will be a service to the numerous researchers and practitioners active in this fascinating field. We allow readers to experience the wonder of peering into a scholarly work of the highest order and seminal significance.

**MAXWELL PRESS**

*To all who may find its contents useful,*
*this book is respectfully dedicated.*

*W. M. H.*

# PREFACE.

THAT this work is placed on the market at all is due principally to the lack of satisfactory, compact reference books treating on the subject in question.

There are many excellent books of reference which treat the subject from a theoretical standpoint and deal largely with the growth and development of the internal-combustion engine. Many of these books, however, have not been brought down to date and, while beyond reproach as exponents of theory, fall far short in the matter of present practice and modern design. It would be well to supplement the use of this book with any one of several works on the gas engine, in order that the mathematical side of the subject may not be slighted. Works by Clerk, Hutton, and Donkin are particularly available along these lines.

A complete knowledge of thermodynamics is invaluable for the perfect understanding of the theory of internal-combustion engines, one of the best text-books on this subject being "Thermodynamics, Heat Motors and Refrigerating Machines," by De Volson Wood.

However, it has been the aim of this work to eliminate, as far as practicable, the more involved mathematical formulas and to confine the matter contained to the more practical and applied phase of the subject. In the chapter on "Compression" several thermodynamic formulas have been used to prove the relation of the compression to the thermal efficiency; these formulas, however, have no immediate bearing, except in a general way, on the problems of actual design and operation, but the formula $PV^n = C$, by far the most important formula used in the actual designing, is found and derived in this chapter, and its discussion is taken up in the following chapter on "The Indicator Card."

For practical information and data contained in this work the author is indebted, to a large extent, to different manufacturers who have placed the result of practical tests at his disposal. It has been the intention to use only that information which appeared most reliable and in keeping with actual practice.

In the matter of design, average practice has been considered, and while the formulas given should not by any means be taken to give results in keeping with each and every engine on the market, their use will insure results closely in keeping with the average.

The tables and formulas herein contained should fill all average requirements, either for the designer or the operator, and while neither original nor compiled especially for gas-engine practice (they may be found in any standard handbook), the fact that they may be found here assembled should be of advantage.

It is the desire of the author in issuing this book that it may find a place for itself and fulfill, in part at least, his intentions.

WILLIAM M. HOGLE.

TOLEDO, OHIO, December, 1908.

## CONTENTS.

### INTRODUCTORY.

### Chapter IV. — Comparison of the Cycles.



### Chapter V. — Practical Operation.



*Care of Engine.*



*Troubles and Remedies.*

## Chapter VI. — Starting Devices.



## Chapter VII. — Carburettors, Vaporizers, and Injectors.



## Chapter VIII. — Producers.

## CHAPTER IX. — FUELS AND COMBUSTION.



## CHAPTER X. — COMPRESSION.

## CHAPTER XI. — THE INDICATOR CARD.



## CHAPTER XII. — GENERAL DIMENSIONS.



## CHAPTER XIII. — THE CAM MECHANISM.



### *The Reduction Gearing.*

CHAPTER XIV. — THE VALVES AND PORTS.



CHAPTER XV. — THE CYLINDER.



CHAPTER XVI. — THE FLYWHEEL.



CHAPTER XVII. — THE FRAME.

### Chapter XVIII. — Engine Foundations.



### Chapter XIX. — The Crank Shaft and Reciprocating Parts.



*The Piston, Wrist Pin, and Piston Rings.*



### Chapter XX. — Governing Devices.



### Chapter XXI. — Ignition.

## Chapter XXII. — Engine Testing.



## Chapter XXIII. — Report of Tests.

MISCELLANEOUS.

# INTERNAL COMBUSTION ENGINES.

## INTRODUCTORY.

THE internal-combustion engine, as we have it in use to-day, is the result of more than two hundred and twenty-five years of experiment, during the greater part of which time, however, no advance was made over the original device produced in 1678. In that year the Abbé de Hautefeuille used the explosive force of gunpowder as a motive power, deriving his work in exactly the same way as the modern internal-combustion engine, by the expansion and subsequent cooling of a volume of heated gas.

Two years later, in 1680, Huygens, a Dutch savant, published a work describing an apparatus, suitably arranged with cylinder and valves, in which the explosion of gunpowder was made to force a volume of heated air into the cylinder, after which, the valve having closed, the gas became cool and soon fell to a pressure less than that of the atmosphere, causing the piston to be forced down by the excess atmospheric pressure. The apparatus as well as the operation was exceedingly crude and no very flattering results were obtained from its use.

Later, about the year 1690, Papin continued the experiments of Huygens, attempting to find a substitute for the gunpowder, the operation of which was very uncertain. His experiments seemed to show very conclusively that the condensation of steam was the most suitable method of producing the vacuum required for the operation of the Huygens motor, and the process was used to some extent in the operation of pumping engines.

Shortly after this time the discoveries of Watt turning

the attention of the public to the steam engine, the development along the lines of internal combustion ceased, and it was not until about the year 1791 that any suggestions were made which were improvements on the engine of Abbé de Hautefeuille. In this year an English inventor, by name John Barber, took out a patent on the use of a mixture of hydrocarbon gas and air in an "exploder."

In 1794 this patent was followed by one covering the production of an explosive vapor by means of a liquid and air. This patent was also taken out by an English inventor named Robert Street.

In the year 1799 Philip Lebon, of Brachay, France, took out a patent on the principle as well as the construction of an engine using the explosion of coal gas as motive power. This inventor also took out patents on a pump for the compression of the explosive mixture and a machine, operated by the engine, for the production of an electric spark for igniting the charge.

The career of this inventor terminating abruptly shortly after this time, and before he had developed his inventions, closed what might have been an epoch-marking period in gas-engine development.

From 1799 until 1860, in which year the first practically successful engine was designed and built, several different schemes were advanced. One brought out by Wright in the year 1833 was very well developed from a theoretical standpoint, a governor being used in connection to vary the mixture of gas to make it proportional to the work being done and to regulate the compression of the charge.

A double-acting engine produced by Johnston, and devised for the use of hydrogen and oxygen, two parts of the former to one part of the latter, was somewhat unique in its operation, and had it not been for the cost of the fuel would doubtless have been used quite extensively. The hydrogen being exploded, formed with oxygen a water vapor which on being cooled was precipitated and a partial vacuum formed, the unbalanced force of the atmospheric pressure then acting during the return stroke of the piston.

In 1838 Barnett took out a patent covering substantially the same ground as did that of Lebon, two pumps being used to compress separately the gas and air and then force them into the cylinder. The explosion was produced by means of the so-called Barnett ignition cock, later described in Chapter XXIII on "Ignition."

The use of the magneto as medium for producing the sparking current was suggested by Stephard in 1850.

In the year 1857 Barsanti and Matteucci devised a motor with a very long cylinder fitted with a piston to which a rack meshing with a spur gear on the fly-wheel shaft was attached. On the explosion stroke a pawl allowed the rack to run freely, but on the return stroke the pawl engaged and the rack caused the spur gear and shaft to revolve. The explosion of the charge drove the piston upwards in the cylinder, and its inertia caused it to pass the point where the internal pressure was equal to the atmosphere and in consequence a vacuum was formed. The cooling of the exploded charge increased this vacuum, with the result that the piston was forced down with considerable force.

In 1858 an engine was devised by Degrand in which the gases were compressed in the cylinder, but because of mechanical difficulties it did not meet with any success, although the idea was a forerunner of the engine of the present day.

The appearance of the Lenoir motor in 1860 marked an epoch in gas-engine construction, as it was the first engine capable of comparatively regular and efficient work. The machine was constructed along the lines of a double-acting steam engine, the ignition was obtained by means of a primary battery and Ruhmkorff coil producing a jump spark, and altogether it was a very decided advance over all existing forms of gas engines up to that time. But the Lenoir engine was uneconomical, requiring about 100 cu. ft. of gas per hp.-hr. and four times as much water for cooling as was used in a steam engine of like power. The great heat in the cylinder required that the piston be kept flooded with oil. In view of these several difficulties the Lenoir engine disappeared in a very short time, but not before it had stirred the minds of the inventors to renewed activity along the lines of the internal-combustion engine.

In the same year, 1860, Hugon introduced a motor in which he attempted to keep down the temperature of the cylinder by means of the injection of a spray of water. This engine was more economical in the consumption of gas, requiring a trifle more than 80 cu. ft. of gas per hp.-hr., and the temperature of the exhaust gases was appreciably diminished.

Several other ideas were advanced about this time, all of them being either of minor importance or repetitions of previous attempts. In the year 1862 M. Beau de Rochas took out a method patent setting forth, theoretically, the best working conditions for an internal-combustion engine. His cycle of operations was in all respects the same as that in use at the present day in the so-called Otto-cycle engines. The following propositions were embodied in his patent:

1. The largest cylinder capacity with the smallest circumferential surface.
2. Maximum piston speed.
3. Greatest possible expansion.
4. Greatest pressure at beginning of working stroke.

While the honor of promulgating the theory belongs, beyond a doubt, to M. de Rochas, he did not in his patent set forth any means for producing the theoretical proposition in practice, and, owing to irregularity in the proceedings, his patent became public property soon after the application was filed, but not until 1878 was attention again called to it. In that year the Otto gas engine, substantially as it now appears, was first placed on the market. Previous to this time, about the year 1872, Otto, in connection with Langen, placed on the market the so-called Otto and Langen engine, of which, due to its comparatively economical operation, they were enabled to sell quite a large number, notwithstanding the fact that it was of the free-piston type and exceedingly noisy in its operation. Its gas consumption was about 26 cu. ft. per hp.-hr., and the cost of energy produced was somewhat less than with the existing steam engines.

Continuing his experiment, Otto, in 1878, produced and placed on the market the first four-cycle engine operating on the Beau de Rochas cycle but commonly known as the Otto cycle. This engine was almost immediately adopted as the standard type of internal-combustion motor, the perfection of which has been the problem of designers.

In 1879 a modification of this engine was produced by Dugald Clerk and formed the basis for present-day two-cycle engine practice. In the Clerk engine the charge was compressed and exploded once every revolution, as against one explosion every two revolutions in the engines of the Otto type.

Since the year 1880 several motors of greater or less value have been placed on the market, but without exception they have disappeared, and at the present time the engines of the four-cycle and two-cycle types, with greater or less modifications, hold the field.

# CHAPTER I.

## THE BEAU DE ROCHAS CYCLE.

REFERENCE to Fig. 1 will explain fully the Beau de Rochas or Otto cycle, the four strokes of which are designated as follows:

1. Suction; 2. Compression; 3. Explosion; 4. Exhaust.

The four strokes as above enumerated form the basis for the term "four-cycle" as applied to gas-engine practice.

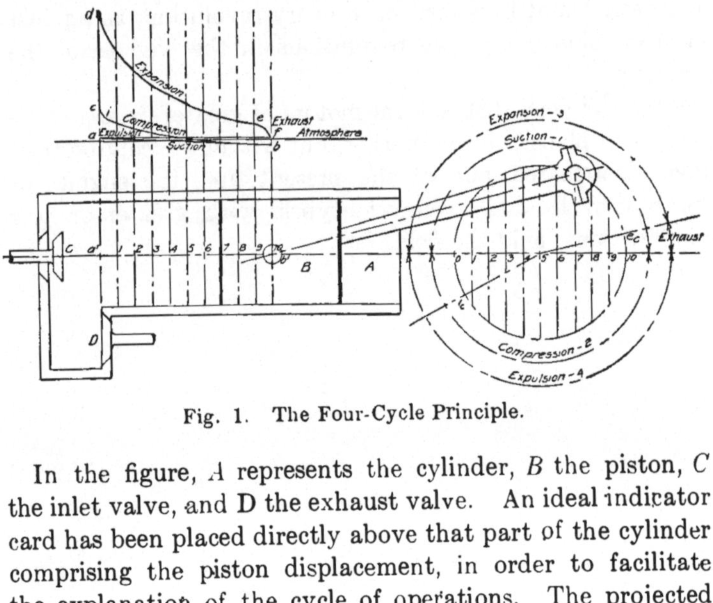


Fig. 1. The Four-Cycle Principle.

In the figure, *A* represents the cylinder, *B* the piston, *C* the inlet valve, and D the exhaust valve. An ideal indicator card has been placed directly above that part of the cylinder comprising the piston displacement, in order to facilitate the explanation of the cycle of operations. The projected path of the crank pin has been divided into ten equal spaces, as has also the path of the piston. From this latter set of points ordinates have been erected and the indicator card drawn. On the indicator card reference points have been indicated by small letters; corresponding points on the crank circle are indicated by the same letters with the subscript *c*. Outside the crank circle two larger circles have been drawn on which the cycle of operations is shown. The inner circle represents the suction and compression strokes and the outer one the explosion and exhaust.

On the suction stroke the port, *C*, opens as the piston travels from *a′* to *b′*, as shown on the indicator diagram from *a* to *b*. Unless valves *C* and D are sufficiently large, point, *a*, on the suction line would tend to be slightly above atmosphere and point, *b*, would tend to be slightly below, due to the resistance of the ports to satisfactorily supply the charge and remove the products of combustion. Especially would this be true on high-speed engines. These results are generally present, to a greater or less extent, in most engines placed on the market. It is obvious that in the different designs of engines the proportion of the ports and valves will vary, and for this reason these defects, in the card, are quite pronounced in some, while in others the suction and exhaust lines will be found to follow very closely the atmospheric line. The carburettor may also be a factor in determining the suction line of the card, as it is apparent that one with a capacity too small for the cylinder displacement cannot supply a full charge under all conditions.

It would then appear that a suction line whose initial point, *a*, was above and whose terminal point, *b*, was below the atmosphere, would indicate that either the engine valves were too small to do their work in a satisfactory manner or that the carburettor was not sufficiently large.

On the compression stroke the piston moves from *b′* back toward *a′*, as shown on the indicator diagram, from *b* back to an indefinite point *i* on the compression curve, *bc*, or to $i_c$ on the crank circle, at which point the compressed charge is ignited, the pressure rising rapidly to the maximum, which should be just as the engine passes dead center at *a′*. As the combustion of the gases is more or less slow, the ignition point is, of necessity, given a positive lead in order to obtain this result. As the velocity of the piston increases it is obvious that the igniter lead must be increased in order to give the charge sufficient time to burn and reach its maximum pressure at dead center. If a card shows a shape as at *d*, Fig. 2, it would indicate that the ignition was not sufficiently advanced, in consequence of which the maximum pressure is not reached until the engine has passed its dead center. The expansion line being nearly an adiabatic, a loss of energy necessarily results. The pressure at release is correspondingly higher, as shown by comparison of the dotted expansion curve of Fig. 2 with that shown in full line, but the mean ordinate is decreased, owing to the fact

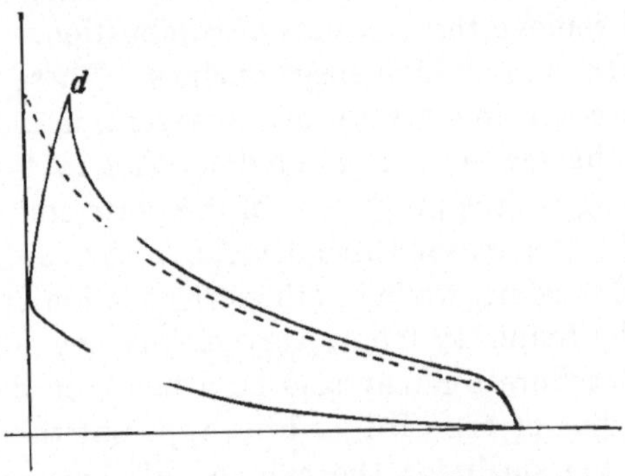


Fig. 2. Card showing Late Ignition.

that the maximum pressure is not developed until after the piston is in the working stroke, and as the ordinates, diminished in length by this late ignition, appear in that portion of the stroke where the pressure is highest, they affect the value of the mean ordinate most.

On the expansion stroke the piston again moves from $a'$ toward $b'$, or, as shown on the indicator diagram, from $d$ to $e$, and on the crank circle the pin moves from 0 to $e_c$, at which point the exhaust port opens, allowing the expansion curve, which up to that point, according to Thurston, is nearly an adiabatic, to drop off quite suddenly nearly to atmosphere. In effect, then, the exhaust takes place from $e$ to $f$. The location of the release, $e$, varies considerably in different makes of engines, the theoretically correct point being so located that the cylinder pressure would fall very nearly to atmosphere just as the engine passed the forward dead center in order that there should be no great amount of back pressure at the beginning of the expulsion stroke; on the other hand, the release must not take place too early in the expansion stroke, as this would seriously impair the power as well as the heat efficiency of the engine.

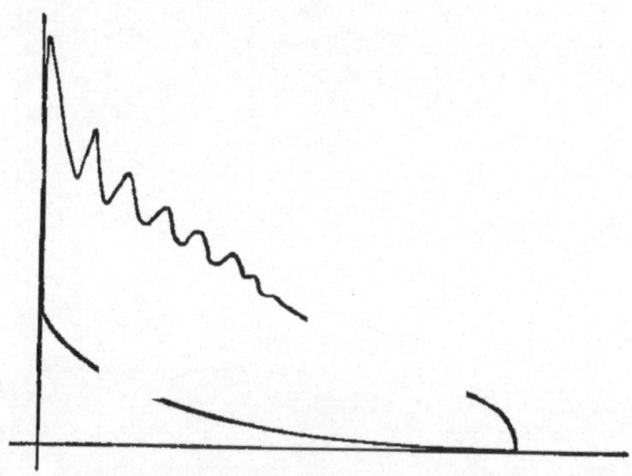

Fig. 3. Card showing Inertia Effect of Governor Spring.

Changing the point of release may be made the basis of one type of governing device (see Chapter XXII).

If the effective area of the exhaust valve is not amply large, then point, *f*, will not be down to atmosphere, and on the expulsion stroke, as the piston again moves from *b′* back to *a′*, the line of the card, *fa*, may be slightly above atmosphere all the way. In well-designed engines these effects are seldom encountered to an appreciable extent, but the possibility of their occurrence makes it necessary to guard against conditions that would tend to produce these results.

This completes the theoretical cycle of operations in the four-cycle engines. In practice the gas-engine card is much distorted, owing to the sudden variations in pressure which take place. The explosion line will be found to run higher in some cases than the maximum theoretical pressure of the burning gases, due to the inertia of the indicator parts. This trouble may be in part remedied by the use of a heavier spring. When the explosion pressure as indicated by the card is too high, the expansion line will assume a ragged outline, due to the harmonic motion of the spring in overcoming the effect of the inertia. This ragged appearance may continue until well into the next compression stroke.

# CHAPTER II.

## THE CLERK CYCLE.

The Clerk cycle engine, more commonly known as the two-cycle engine, as previously mentioned, was first introduced by Dugald Clerk about the year 1879, and was the first of the compression motors built, receiving an impulse every revolution. In the Clerk motor its inventor introduced the charge into the cylinder under compression, as is done in the present types of two-cycle engines. Instead of obtaining his primary compression in the crank case, as is the modern practice, he used an auxiliary pump. The exhaust ports were arranged in the cylinder wall, being uncovered by the piston on its downward stroke, the inrush of the compressed charge expelling the products of combustion in practically the same manner as this is accomplished in the later two-cycle engines. This type of motor was composed of two cylinders, one the power cylinder and the other the primary compression cylinder. The primary cylinder communicated with the explosion cylinder at the top of the compression space, and herein differed from the present type, in which the charge is introduced through a port very nearly in line with the exhaust port; it being uncovered, with the exhaust port, by the downward stroke of the piston.

The theory of the Clerk engine is the same as that of the Day two-cycle (i.e. introduction of the charge under compression), but the added number of parts with the consequent addition of weight and complexity is a feature which made this type of engine impracticable for general use, more especially where a light motor is required; and as nothing is gained, in its construction, over the Day type, the balance is all against it. Nevertheless the honor of first describing the two-cycle principle must be accorded to Clerk and the practical improvements and application to later inventors.

## The Robson Engine.

A forerunner of the Day type of motor was the Robson engine, manufactured by Messrs. Tangye under Robson's patent. In this engine the cylinder was closed at both ends and used a piston rod. The forward end of the cylinder was used for obtaining the primary compression, the charge being drawn in on the suction stroke and compressed during the greater part of the explosion stroke, and the gas thus compressed was forced into the power cylinder through an automatic lift valve, which operated when the piston was fully out and the exhaust valve wide open. This charge was then compressed by the return stroke of the piston and exploded as in the ordinary two-cycle motors. Two valves were necessary in this engine, an automatic valve for admitting the charge to the primary compression chamber and one opening from this space into the power cylinder. The engine was of rather neat design and not nearly as cumbersome as the Clerk production.

## The Stockport Engine.

This engine was exactly similar in principle to the Robson engine, but the forward end of the cylinder was not utilized to obtain the primary compression. The engine was similar in design to the two-cylinder opposed motors of the present day, one cylinder being used for the primary compression, while the other was used for the power cylinder.

Several other types of engines were evolved and placed on the market about this time, but the most of them were either too cumbersome or too complicated to meet with general use, and the Day cycle proper has almost, if not quite, displaced these earlier types.

## The Day Engine.

As the Day engine is the analogue, in two-cycle construction, of the Otto engine in four-cycle design, a complete description of it will be given. In the description of the Day engine and cycle we are describing, in principle, the original Clerk idea.

Referring to Fig. 4, $A$ represents the cylinder, $B$ the piston, $C$ the inlet valve, and D the connection between the crank case, in this instance the primary compression space, and the power cylinder. An ideal indicator card has been placed directly above that part of the cylinder comprising the piston displacement, in order to facilitate the explanation of the cycle of operations. The projected path of the crank

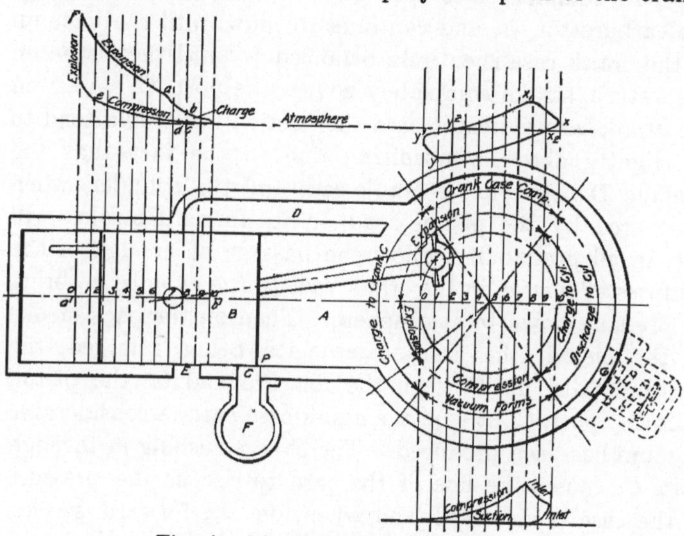


**Fig. 4. The Two-Cycle Principle.**

pin has been divided into equal spaces, as has also the path of the piston. From the latter set of points ordinates have been erected and the indicator card drawn. On the indicator card reference points are indicated by small letters; corresponding points on the crank circle are referred to by the same letter with the subscript $c$. Outside the crank circle another larger circle is drawn, on which the cycle of operations in the cylinder is shown; on a still larger circle the cycle of operations in the crank case is shown. Directly above the projected path of the crank pin the crank-case indicator card has been constructed, the points of reference being indicated by the letters $x$ and $y$ with suitable subscripts.

On the suction stroke the piston travels from $b'$ toward $a'$, or, as shown on the crank-case indicator card, from $x$ toward $y$. Just as the forward end of the piston passes point $P$, the inlet valve, $C$, starts to open. A vacuum has been produced in the crank case up to this point, by the piston traveling upwards in the cylinder, and in consequence when this port commences to open a charge of gas rushes in from the carburettor, $F$, and continues to flow until the vacuum in the crank case is entirely balanced or until the piston on the return stroke completely covers the port, $C$. At $x$ on the crank-case indicator card the suction line would tend to be slightly above atmosphere, due to the fact that the opening D between the crank case and the cylinder, when uncovered by the piston head on its forward stroke, will not, in all probability, allow the passage of enough of the compressed charge in the crank case to the cylinder to bring its pressure down to atmosphere. Then on the back stroke of the piston, the crank-case space being enlarged, the pressure falls, until, when the forward end of the piston uncovers port, $C$, at $y$ on the crank-case card, a considerable vacuum has been produced. The charge rushing in through port, $C$, causes the line of the card to rise as the pressure in the case is increased, and when, on the forward stroke, the port, $C$, is again completely covered, the card should show a pressure of about atmosphere as at $y_2$. If port, $C$, is too small, this will not be the case, but the point $y_2$ will still show a slight vacuum, which will necessarily affect the maximum crank-case compression at $x_1$ when the piston head uncovers port, D. From $y_2$ to $x_1$ on the card the pressure should rise regularly until at $x_1$ the pressure should be maximum, which should be from 6 to 10 lb. per sq. in., in no case less than 5 lb. per sq. in. At this point, port D being opened by the piston, the pressure line commences to fall, and continues to do so until the pressure is equalized on both sides of the piston, or until the piston on its return stroke again closes valve, D (shown at point, $x_2$, on the indicator card).

Since the displacement of the piston is the same on the

cylinder end as on the crank-case end, the charge taken into the crank case on the suction stroke should exactly fill the cylinder space, and the pressure at $x_1$ should be equal, in pounds, to the vacuum at $y_1$. But it is practically impossible to so proportion the ports that they will produce these results at all speeds, unless they are made abnormally large. This is especially true in the three-port type of engine here described. The charge introduced into the crank case on the suction stroke does not vary with the speed at which the engine is running, as the vacuum produced will not vary to any considerable extent except as the increased speed of the piston gives less time for leakage of air into the crank case before the inlet valve opens, consequently on slow speeds the port being open longer allows a larger charge to rush in. This accounts for the inability of some two-cycle motors to run at high speeds and deliver power in proportion.

As noted, these conditions are more noticeable in the valveless engine than in the engines using a poppet valve for the inlet port, as indicated at *G*, Fig. 4, in which case inlet port, *C*, is not used. In this construction it is always possible to get a full crank-case charge with a correspondingly higher pressure at $x_1$, but we are still confronted with the difficulty of making port D large enough to give a full charge at all speeds, but, as pressure $x$ is greater, the flow will be somewhat more rapid. It is probable, however, that this advantage is sufficient to make the balance favor the engine with the valve over the more simple three-port engine; in fact many manufacturers have tried and discarded the three-port type. The crank-case card for the two-port type is shown in Fig. 4 below the crank case.

Having followed through the cycle of operations in the crank case, let us look at the operations taking place in the cylinder. On the cylinder indicator card let *a* represent the point of opening of the exhaust port, *E*, *b* the opening point of inlet port, D, *c* the point of closing for inlet port, D, and *d* the point of closing for exhaust port, *E*.

On the forward stroke of the piston *B*, the charge having

been exploded, at point *a* on the card the piston head commences to uncover exhaust port, *E*. At this point, then, the pressure of the expansion line falls off rapidly, forming some such a break in the card as is indicated. At point, *B*, on the indicator card the port from the crank case into the cylinder commences to open; at this point, *b*, it is obvious that the pressure of the exhaust gases in the cylinder should have fallen to such an extent as to make the cylinder pressure less than that of the crank case, otherwise the cylinder will exhaust back into the crank case and cause what is known as back firing, with consequent loss of power. For this very reason the exhaust port is given a lead over the inlet port, varying to some extent in different two-cycle designs (see Chapter XVI). From *b* to *c* on the indicator card the crank-case charge rushes through port, D, to fill the cylinder space. A baffle plate, *G*, is placed on the piston head to deflect the incoming charge to the top of the cylinder, so that it may more effectively force out the burned gases without being exhausted itself. At point, *d*, the exhaust port closes and the cylinder compression commences. The charge is compressed from point, *d*, up to the point of ignition, which of course is varied for different speeds by the spark-timing device. From *e* to *f* the explosion takes place and the expansion follows from *f* back to *a*.

In two-cycle work the following points should be striven for: 1. Quick exhaust, with consequent large exhaust port. 2. Unrestricted exhaust port. 3. Crank-case compression high enough to make the primary pressure greater than the cylinder pressure when the inlet port opens. 4. As large inlet ports as possible. 5. As long a compression and expansion line as possible. A little study of the card will serve to show that some of these points must necessarily be sacrificed, to a greater or less extent, in order to attain the others. The expansion line must be shortened in order to give sufficient exhaust period, and a shortening of the expansion line produces a corresponding shortening of the compression line. The most satisfactory arrangement of these several points will be taken up later in the design.

# CHAPTER III.

## THE DIESEL MOTOR.

WHILE the cycles of Beau de Rochas and Clerk cover, in the broadest sense, all modern practice along the line of internal-combustion engines, at least one other engine has been produced and placed on the market which, while it utilizes the four-cycle principle as a basis for its construction, has made use of a different method for obtaining its fuel charge. This engine was the invention of Rudolph Diesel, a German scientist and inventor. The cycle requires two revolutions of the crank shaft for its completion; the first outward stroke of the piston draws into the cylinder a charge of pure air instead of combustible mixture, and on the following return stroke of the piston this charge is compressed to a pressure of about 500 lb. per sq. in., at which pressure its temperature is sufficient to ignite any form of crude or refined petroleum. When the piston has reached the top of the compression stroke the fuel valve opens and a charge of vaporized fuel is injected into the incandescent cylinder by means of air compressed to about 800 lb. per sq. in., *but cooled before it reaches the fuel valve.* It is the intention to maintain the temperature of combustion constant at the temperature of compression, thus allowing the pressure to fall in accordance with the laws for expansion at constant temperature, or, in other words, to make the combustion curve as nearly as possible an isothermal, this being the curve of maximum economy for an internal-combustion engine. It is obviously impossible to obtain this result except when the engine is running under very nearly normal conditions. When the engine is overloaded and the amount of fuel is increased by means of the governor, the temperature will rise above that of compression, in order that the mean effective pressure may be higher, while with a light load the temperature and pressure will fall, due to the quantity of fuel being less.

After the period of fuel injection, which comprises about 10 per cent of the working stroke, is completed, the fuel valve closes and the ignited charge expands until 90 per cent of the working stroke has been completed, at which point the exhaust valve opens in order to relieve the pressure before the expulsion stroke commences. The manufacturers claim their pressure at exhaust to be about 35 lb. per sq. in. for normal load, which pressure would necessarily be increased or diminished as the engine was operated at overload or running light.

The fourth stroke in the cycle is the expulsion stroke during which the piston, traveling upwards with the exhaust valve open, ejects the burned charge.

In the Diesel cycle there is no opportunity whatsoever for premature explosion, since the fuel is not injected until the beginning of the working stroke. The high compression and correspondingly small compression space, about 7 per cent of the cylinder volume, make it possible to eject nearly all of the burned gases and to secure a charge of almost pure air to support the combustion during the working stroke. The fuel economy, with the theoretical conditions attained, would necessarily be high, and actual results seem to prove this to be the case. There is some question, however, as to the wear and tear resulting from the heavy parts made necessary by the long-sustained high compression. This high compression causes the temperature in the cylinder to approximate for a much longer period the temperature of combustion, but as this temperature of combustion is much lower than in most internal-combustion engines, it is probably true that the parts subjected to this heat are not damaged to any appreciable extent, and it is doubtful if the long-sustained compression would be more harmful than the suddenly applied pressure induced in engines operating on the Otto principle.

Fig. 5 will explain fully the cycle of operations taking

place in the Diesel engine. The same arrangement of diagram has been followed in this as in the previously described cycles.

*xa* represents the clearance.

*ab* represents the stroke.

The stroke is divided into ten parts by the ordinates 0–10, as shown. In the engine diagram $v_1$, $v_2$, and $v_3$ are respectively the inlet, exhaust and fuel valves. On the first for-

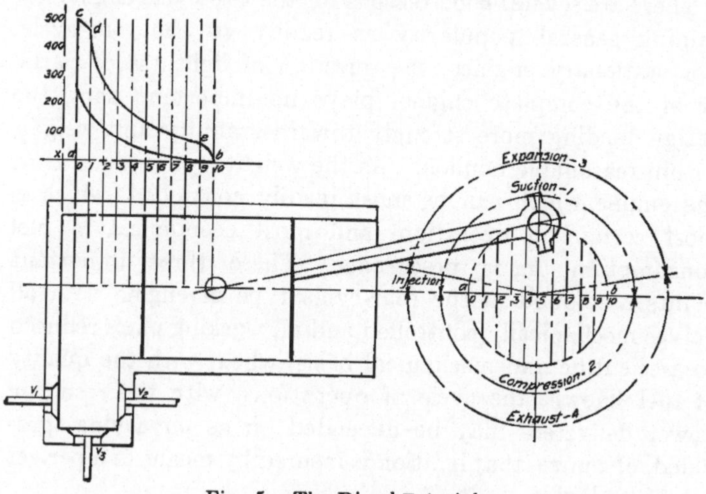


Fig. 5. The Diesel Principle.

ward stroke of the piston, from *a* to *b*, occurs the suction of pure air, valve, $v_1$, being open. On the backward stroke of the piston the air is compressed, following the curve *bc* on the card, all the valves being closed. On the second forward stroke of the piston, from *cd* on the card, the fuel valve, $v_3$ opening, the fuel is injected. The curve *cd*, as previously described, is an approximate isothermal; at *d* a break occurs, as the fuel valve is closed; and from *d* to *e*, or during about 80 per cent of the working stroke, expansion takes place. At *e*, the exhaust valve, $v_2$, being opened, the pressure falls and reaches atmosphere at *b*; and from *b* to *a* during the expulsion stroke, on the second backward stroke of the piston, the products of combustion are discharged.

# CHAPTER IV.

## COMPARISON OF THE CYCLES.

The four-cycle engine has found most favor with the general public, and in consequence has been most widely manufactured. In stationary engine practice this design is in almost universal use, although there are some two-cycle engines in use for this class of work, and it seems that they are gaining some headway.

There are several good reasons for the two-cycle engine not gaining general popularity as readily as the four-cycle. For stationary engines, the question of lightness of parts, or of the complete engine, plays no important part, the design tending more strongly toward stability and weight, within reasonable bounds. As the weight is no inducement, the engine which can be most readily controlled, which is most certain in operation, and most economical in fuel consumption, gains precedence. These three important points are found in the four-cycle type of engine with all valves mechanically controlled and all working parts reduced to a scientific and mechanical basis, when, with the quality of fuel known, the cycle of operations, with the resulting power delivered, may be depended on as unvarying, provided, of course, that ignition is insured by means of a perfect sparking device.

In automobile engines the four-cycle type, while not universally adopted, is used in the very large majority of cases, but in a somewhat modified form. Here the different manufacturers have striven to create a machine with the greatest power and the least weight; in other words, they try to make the weight of the engine per horsepower as low as possible. To accomplish this many manufacturers use the suction inlet valve, but not to as large an extent as formerly, the mechanically controlled inlet and exhaust being looked upon with more favor at the present time, due to the fact that the suction inlet has not, in all cases, given universal satisfaction, and in fact was quite unsatisfactory in many instances.

In marine-engine practice probably about an equal number of two and four cycle engines are manufactured and sold. The two-cycle type is more satisfactory for marine use than for any other purpose, as the necessary amount of cooling water is more readily available. Since in the two-cycle type of motor, an impulse or explosion occurs at every revolution, it naturally follows that the cylinder heats up more rapidly than in the four-cycle type, in which the impulse, occurring only once every other revolution, gives the cylinder more chance to cool.

While the two-cycle engine grows hotter, due to the more frequent explosions, it should produce more work and steadier power for the same reason. This will be found to be the case, if the ports are properly proportioned; but no two-cycle engine ever built could produce twice as much work for the same number of revolutions, at all speeds, stroke and bore being the same, as a four-cycle engine. The reason for this is found when one tries to proportion the ports for a gas speed of 100 ft. per sec. for the inlet and 90 ft. per sec. for the exhaust and finds that the ports must necessarily be quite large to admit a full charge at 800 rev. per min. and under the most favorable conditions, that is, with the cylinder completely scavenged of the previous charge and the carburettor sufficiently large and properly adjusted to allow a full charge to enter the crank case on the suction stroke. In the majority of two-cycle engines the ports are made much smaller than they should be, even for their nominal speed, with the result that when the speed is increased the charge is very greatly diminished. At speeds where the four-cycle engine would obtain a full charge of gas it is probable that a two-cycle motor does not obtain much more than half a charge, and as the speed is further increased, even less than that amount. It is doubtful if the average two-cycle motor does more than 30 per cent more work than a four-cycle motor of the same size, both operating under most favorable conditions.

The two-cycle motor is more wasteful of fuel than the four-cycle type, and trouble with crank-case explosions is frequent, due to the explosion following back into the crank case when the crank-case compression is lower than the cylinder pressure as the inlet port opens. This may be obviated, in a large measure, by placing a screen baffle plate in the inlet passage; the screening acts on the principle of a Davy miner's lamp. The baffle plate must be a close fit, however, to be effective, and must be inclined in the passage, so that the meshes, through which the gas passes, may be equal in area to that of the port itself.

The Diesel motor is gaining some favor as a prime mover in power plants. Its performances, as far as known, are excellent, its manufacturers claiming the exceptionally high efficiency of 38 per cent. The engines are accompanied by a guarantee, as to fuel consumption per horsepower-hour, good for one year from the date of installation. The question of first cost and the necessarily heavy parts possibly inducing an extraordinary amount of wear, are, as a general rule, the most serious obstacles to prospective purchasers. However, reports from plants in actual operation tend to show that the manufacturer's guarantee is none too high. In fact, some operators claim their fuel consumption to be below the manufacturer's guarantee.

The municipal lighting and waterworks plant of Bryan, Ohio, report that their motors have given not the slightest trouble from regulation or wear, and that they have made runs of a month at a time without stopping.*

* The author is indebted to Mr. S. L. Folk of Bryan, Ohio, for the information in regard to the practical operation of the Diesel motor.

# CHAPTER V.

## PRACTICAL OPERATION.

*Starting a Stationary Engine.*—Starting a gas engine is in most cases a simple operation, if a few rules are remembered.

A gas engine will not start without it obtains enough initial power from some outside source to enable it to commence its cycle of operations.

The mixture of fuel and air must be neither too rich nor too poor in fuel, for if either of these conditions obtain, an explosive mixture will not result.

In starting a stationary engine by hand, or in fact any other way, the ignition should be given a negative lead, or, in other words, the sparking point should be past the upper dead center in the direction in which the engine is running. Failure to note this important point will result in back firing, with more or less disastrous results.

Compression should be relieved and the load thrown off, unless a powerful starting device is used, as, for instance, the compressed-air system.

Always be sure that the engine is well oiled, and the oil cups are filled.

See that the ignition apparatus is in good order and the sparking points clean, if electrical ignition be used. If the fuel used is clean and burns without producing a large amount of soot and crust, the sparking points will remain clean much longer than if a dirty fuel is used. Use a cylinder oil with a high flashing point, in order to obtain the best results, as an oil which flashes at a low temperature will assist, very materially, in fouling a cylinder. A foul sparking device cannot be made to yield good results.

Be sure that all wiring connections are close and clean and that the batteries and coils are in good working order.

If a hot tube igniter is used, bring the tube to a cherry-red heat and adjust the flame to maintain this temperature.

If the engine has a starting cam (a double cam acting on the exhaust valve and serving the same purpose as a relief cock), it should be thrown into starting position. If the engine is provided with a relief cock, instead of the cam, open the cock.

Set the igniter to the proper starting position and open the valve in the gas supply pipe about one-quarter full, or possibly a little more. Now give the engine an impulse, and, as soon as it begins its cycle of operations, commence slowly to open the gas valve and continue until the engine is getting its maximum supply and is running at its regular speed. Do not open the valve too rapidly, or the engine will get too much gas, and in consequence too rich a mixture; it will soon slow down and stop. While the valve is being opened, the relief valve mechanism may be thrown out of gear, or the relief cock closed, and the spark or ignition device advanced to running position, unless a governor acting on the igniter mechanism is used.

When the engine is well started, the load may be thrown on and the water turned into the water jacket until the discharge water is at a temperature of from 160 deg. fahr. to 180 deg. fahr., for stationary engines of low compression. For high-compression engines, a somewhat lower temperature is necessary.

If the engine is provided with a starting device, allow it to make several revolutions before opening the gas valve. (See later chapter on "Starters.")

As soon as the engine is running well, inspect all oil cups and make sure that they are feeding properly.

*Stopping.*—To stop a gas engine turn off the gas valve, and if it is desirable to stop the momentum of the flywheel, a friction brake, in the shape of a plank, may be made to press against its rim, by placing the stick against the floor or other available fulcrum, and prying against the rim. Turn off the oil and water supply and turn off the flame in the hot tube igniter, if one be used, or if electrical ignition is used, turn off the switch. If there is the slightest danger of the jacket water freezing, or if the engine is to be left any length of time, drain the jacket. A little care in this direction will often prevent a crack in the water jacket.

*Starting an Automobile or Marine Engine.*—The same general rules apply in the starting of these machines as in the starting of a stationary engine. However, they are, in nearly all cases, turned over by hand until the cycle of operations is established, and they are always found to be far more erratic in their action than the stationary engine, due largely to the varying mixtures obtained from the carburettor. This mixture, under different atmospheric conditions, varies to such an extent that, where one day a machine may be started with a single cranking, on another day, or even a few minutes later, with the same adjustment of the carburettor, it will not start at all.

Of course trouble in starting may not always be due to the carburettor: the wiring connections may be poor, a battery connection may be broken, the sparking coil may be out of adjustment or the points corroded, or the batteries may be weak. The gasoline, not infrequently, is of poor grade; personally the author has encountered gasoline containing more than 50 per cent of water, although this condition was, of course, extreme. By filling the gasoline tank through a funnel, chamois skin lined, this latter difficulty may be obviated, as gasoline passes through and the water does not.

In starting, the following mode of procedure should be carried out. Retard the spark advancing lever to a point beyond dead center, in the direction which the engine runs, open the gasoline supply valve, if such an arrangement is provided, or, if the machine is supplied with a pump, give it three or four strokes.

See that the clutch or transmission lever is in neutral position, or, if a foot clutch mechanism is used, see that it is thrown out of gear.

Close the sparking circuit, either by inserting the plug or by throwing the switch.

Open the oil supply valves.

Now give the engine a few turns with the starting crank, and, if it fails to start at once, try priming the carburettor; this will usually solve the difficulty; but if it still fails to start, prime the cylinder, through the priming cup usually provided, using care not to use too much gasoline; a few drops is sufficient. Failure to start at once, after these attempts, shows that the cylinder either is getting too rich gas, or is flooded, or that some other part is out of order. (Troubles and remedies are more fully discussed later in this chapter.)

There are two ways of starting a two-cycle marine engine. One way, the engine is turned over the same as a four-cycle until it takes up its operation. The other way, and the one most frequently employed in small engines, is to work the flywheel back and forth to get a charge into the cylinder, then with the spark retarded in the direction in which the engine is to run (a two-cycle engine is reversible), turn the flywheel sharply back against the compression, until it sparks, instead of turning it over dead center.

The same difficulties in starting are found in the two-cycle engine as in the four-cycle type.

When the engine, either automobile or marine, is well started, gradually advance the spark, and open the throttle to running position; as this is a variable quantity in these types of engines, no fixed rule can be given, but do not open out too quickly; give the engine time to "catch up."

Always be sure that the engine circulation is good and that the oil supply is working properly; too much oil is bad, but too little oil is worse.

*Stopping.*—To stop an automobile engine, throw off the spark or close the gas supply, shift the transmission into neutral and apply the transmission brake. After stopping the engine, close the oil supply valves, if a force lubricator is not used, and, if the machine is to be left any length of time, remove the plug from the coil or throw off the switch.

In stopping a marine engine, the propeller acts as a brake,

and it is only necessary to throw off the spark or close the gas supply. The same general rules apply to closing the oil valves and leaving everything about the engine ready to start again.

### Care of Engine.

As a matter of fact, a gas-engine plant requires less attention, by far, than a steam plant of the same size. However, the gas-engine owner or operator should not confuse this statement, or similar statements, into meaning that a gas engine requires no attention and will "run itself" after starting.

A stationary gas engine should have its regular attendant, who, while he need not give his entire attention to the engine, should be depended on to see that it is always in good running condition. A gas engine should always be as clean and as well oiled as a steam engine, and it should always have a sufficient supply of jacket water to maintain a uniform temperature of from 170 deg. fahr. to 180 deg. fahr. It is of importance that the temperature of the cylinder be kept uniform, especially in the case of an engine running electrical machinery, as variations in temperature may be readily detected in the operation of the engine. In order to maintain a uniform temperature, the pressure of the water at the jacket must be kept constant. One of the best ways to accomplish this is to depend on the water pressure of a uniform head of water instead of direct pressure from the circulating pump. This may be accomplished by pumping the water first to an elevated tank and allowing it to circulate from there to the engine and then to the pump, from which it is again pumped to the tank. Suitable means for cooling should be provided, either in the shape of a cooling tower or other device by which the temperature of the water may be lowered as rapidly as possible. It is always advisable to use the cooling water over and over, since, after two or three circulations through the jacket, it will be "broken," that is, the lime or other impurity contained will have been precipitated; frequent renewal of the jacket water will quickly cause a crust to form in the jacket or sediment to lodge at some point, at which place a "hot-spot" will be produced.

If, as is sometimes the case, the exhaust valve is provided with means for circulating the water through it, the water passages should be drilled out as often as once a week in order to insure their remaining open to circulation.

The exact point of ignition should be known, so that, in starting, the engineer may know when to expect the explosion; also the sparking device may, for some reason, become out of adjustment, or it may be necessary to remove it for repairs. With a make-and-break electrical ignition system, this point may be determined as follows: slowly turn the engine over until the igniter snaps, at which point the spark is produced. Now, without moving any part, make corresponding points on the flywheel and frame or on the piston and cylinder, the latter way being most desirable if possible. It is obvious, then, that the engine may, at any time, be turned to its sparking point, even though the igniter is removed; that is, it may be turned to the exact distance from dead center where the ignition occurs; but here, in the four-cycle engine, a difficulty confronts us: we must be sure we are in the *explosion stroke* and not in the *suction stroke.* This may be most readily determined by inspecting the cams. If both cams are down, then both valves are closed, and we are all right; but if the inlet cam is just commencing to raise the valve, we are in the suction stroke and must turn the engine over one complete revolution until the reference marks again correspond, at which point the sparker may be set to snap.

All first-class engines, when they leave the shop, should have their valve and spark positions marked; and these marks, together with printed instructions, should enable any average mechanic to reset the valve or igniter mechanism.

The care of the ignition mechanism is an all-important part in the operation of a gas engine. Electrical devices, if properly cared for, give excellent satisfaction, while if allowed to become dirty or out of adjustment they will give very poor satisfaction. Be sure that the connections are clean and close and that the battery, if one be used, is not allowed to run down. It is good practice to have two sets of cells connected up with a switch, by means of which either set may be thrown into circuit.

If a tube igniter is used, the best material that can be purchased is none too good. Nickel alloy or porcelain withstands the action of the heat and gases best, and, with ordinary care, a tube of either of these substances will last a comparatively long time, while an iron tube needs to be replaced every few days. Several tubes should always be kept readily available so that, in case of accident to one, another may be quickly substituted. The tube should be kept at the very lowest temperature at which the gas will ignite and should never be hotter than a cherry-red. In practice, some gases will be found to inflame more readily than others as the quality is richer or poorer.

The bearings and running parts of a gas engine should be well lubricated with a good grade of machine oil, but the cylinder should be lubricated with a gas-engine cylinder oil of high flashing point, or otherwise the carbonized oil produced will soon choke the passages, prevent the valves from seating, and, becoming incandescent, cause premature ignition and back firing. In any event the exhaust passages should be cleaned occasionally to prevent any possible accumulation from reducing their effective area, thus producing a back pressure and reduction of power. The valves should be frequently examined and ground in with flour, emery and oil, if they leak ever so little. The valve-stem springs should be stiff and strong; if they become weakened it is not always necessary to replace them, but they may be removed and stretched to increase their strength. The few engines that use the suction inlet observe the reverse of this rule, and in their case the valve spring should only be strong enough to properly seat the valve, so that a small vacuum, in the cylinder, will open it quickly. The seat for a suction-inlet valve must always be perfect, since the pressure of the spring is not sufficient to exert any appreciable grinding effect.

Never turn the circulating water into a hot cylinder too rapidly, or the sudden cooling of the walls may cause them to contract, while the piston is still hot and expanded, with the result that the piston sticks and cuts the inner surface of the cylinder. A ring cut, once started in the cylinder, will grow until the compression of the engine is ruined and its power gone. When a cylinder is cut badly it can be repaired only by reboring and providing a new piston and rings.

The governing device should receive frequent attention to prevent its becoming clogged or gummed up with grease and losing its sensitiveness. This is especially true in the case of governors contained in the crank case, where they are in a position to accumulate a great amount of dirty grease. For this reason it would be much better design to place the governor in an apartment by itself, or even to leave it exposed where it may be easily attended to. The hit-and-miss governors act on the gas supply by opening and closing a gas valve; as the engine increases speed beyond a certain limit, the governor catches and closes the valve, or releases it and allows it to close, and it will remain closed until the engine slows down enough to allow the valve mechanism to connect again. If the gas supply valve is not open enough, the engine will not get a charge and impulse the first time the governor connects, and the engine will slow down until the aperture opens wide enough or long enough to allow a charge to enter the cylinder. A hit-and-miss governor, properly adjusted, and with proper opening of the gas valve, should govern the engine very closely.

There are a number of the so called hit-and-miss governors of different design on the market, all acting on the same general principle of closing off the gas supply.

With the hit-and-miss governor, the first impulse received, after the governor connects, is always stronger than the normal, due to the fact that all the hot exhaust gases have been expelled, and the cylinder, in consequence, gets a full charge and a cooler one than usual.

The quality of the mixture should be watched, and may be determined by inspecting the exhaust, which should be almost colorless, what color there is being imparted by the cylinder oil, which will always burn more or less. The nature of the combustion may also be determined by opening the relief cock, if one be provided, and watching the color of the flame, which, for perfect combustion, should be deep blue, bordering on a violet.

The cylinder and piston and the valve stems should be cleaned occasionally with kerosene, and no oil that will gum or carbonize should ever be used on the valve stems.

The crank-pin bearings and the main bearings should be inspected from time to time, and adjusted at the first sign of wear or looseness. The method of impulse, in a gas engine, will loosen and wear the bearings much more rapidly than in a steam engine, and once they start to loosen and the engine commences to pound, the trouble will grow very rapidly.

## Troubles and Remedies.

Trouble in the operation of a gas engine is due more frequently to ignorance in handling than from any fault of the engine itself. Ignorance in handling may also be understood to include careless handling and inattention to small details, which, if given their proper consideration, will assist very materially in the successful operation of an engine.

In the enumeration of the troubles connected with gas or gasoline engine operation, the subject will be treated as a whole, it being understood that the carburettor difficulties apply only to that class of engine in which liquid fuel is used in connection with a carburettor. Nearly all other sources of trouble are common to both engines using liquid and gaseous fuel.

*Engine Fails to Start.*—If the engine will not start, examine the gas valve to see if it has been open too long and allowed too much gas to leak into the cylinder, or if it is open too wide, allowing too full a charge to be taken. If either of these conditions is found to be true, close the valve entirely and turn the engine over once or twice, to clear the cylinder, or until an explosion occurs; then open the valve to starting position and try starting the engine. If this fails, go over the ignition system thoroughly, as described under heading "Spark Weak or Wanting," or, if the hot tube igniter is used, see that the tube is hot enough to ignite the charge.

*Cylinder Flooded.* — Partially close the gasoline supply and turn the engine over enough times to satisfy yourself that all surplus gas has been worked out of the cylinder.

*Carburettor out of Adjustment.* — As every carburettor is different, the engine operator must familiarize himself with his special one and find in what adjustment it produces, on the average, the best results. It is useless to attempt any fine adjustment of the carburettor while the engine is not running, but it may be set to its approximate adjustment, once that is known.

*Spark Weak or Wanting.* — If the spark grows weak, the batteries are probably poor or old. This trouble may be remedied, to a certain extent, by adjusting the points of the coil; for weak batteries the points should be set much closer than when the batteries are strong. The spark may be tested by removing the spark plug and holding it, by means of the insulated wire, against the cylinder; then turn the engine over to see if there is a good fat spark between the points. If the spark is weak and uncertain when exposed to the open air, it will be very much weaker when under cylinder compression, with the probability that there will be no spark at all between the points under these conditions. When the spark plug is taken out, see that the points are set the proper distance apart; the size of this spark gap will vary, to some extent, with the age or strength of the batteries, but 1/32 in. is about right. Be sure that the points are clean and free from soot; to insure this condition, they should be cleaned, from time to time, with gasoline.

If an extra set of cells is carried, throw them into the

circuit and see if the spark is improved. If a good spark is not produced after these trials, look over the wiring connections, through the batteries, and complete the circuit, to make sure that no wire is grounded or contact broken. A well-insulated wire will often ground if covered with oil and grease, and for this reason the wiring should always be kept as clean as possible.

If the wiring connections are all found to be in good condition and neither set of batteries gives a good spark, they are, in all probability, both in need of renewal. There is a possibility, of course, that the coil is poor or that the insulation of the high-tension wire has broken down, but this trouble is not of common occurrence. Open-circuit cells, or cells that polarize rapidly, are not suitable for the rapid work required of a sparking battery.

*Engine Stops.*—If suddenly, in all probability a wiring connection is broken. If it slows down and stops, the cylinder is either overheated, the gasoline low or of poor quality, the mixture is not right, or the spark is too weak to explode the charge every revolution. The engine may be overloaded, or there may be an abnormal amount of friction at some point, due to an overheated bearing or to lack of sufficient or suitable cylinder oil. The inlet or exhaust valve may be obstructed and unable, in consequence, to seat itself properly, thus spoiling the compression and power of the engine. The valve stems may be sticking, with the same result. The relief cock may be partially or wholly open, or the starting cam may be in gear.

If the engine is found to be running hot, ascertain if the circulation is good; there may be an obstruction in the suction line to the pump. If the water seems to be circulating, see if any part of the cylinder seems hotter than the rest; if the engine is vertical, that part of the jacket opposite the water inlet usually heats up first, due to the poor circulation at that point, and the resulting deposit of impurities from the water. If the engine is horizontal, the hot part of the cylinder will usually be found at the bottom, for the same reason. If a "hot-spot" is located, it is usually due to the presence of a deposit; and to improve the circulation, this deposit must be removed. Many engines provide an opening into the jacket at these points and fit the opening with a pipe plug, which may be removed, when necessary, to flush out the jacket. If the engine is so equipped, remove the plug, and, using a bent wire, break up the deposit and then, with the water outlet closed, force water through the jacket until it is clear. If no such plug is provided, and the engine continues to give trouble from overheating, it may be found necessary to drill and tap a hole for a 1-in. pipe plug.

*Ignition Tube Cold.*—If the ignition tube is too cold to fire every charge, then some unburned gas will be discharged into the exhaust passages and explode there. If the tube does not fire the charge frequently enough to keep up the cycle of operations, the engine will stop.

*Mixture too Rich.*—This condition usually results in explosions in the exhaust passages, or in stopping of the engine.

*Back Firing.*—When this condition exists, the charge fires back, in the compression stroke, against the direction in which the engine is running. Back firing may be due to any one of a number of causes; the compression may be too high, but this should not result in back firing, except at low speeds, as on high speed the charge should be ignited considerably ahead of dead center in order to allow the gas to expand to its maximum pressure by the time dead center is reached; it is doubtful if, unless the pressure were abnormally high, the charge would ignite before this critical point was reached. There would be more likelihood of the condition being encountered in gases of low ignition temperature, as gasoline vapor. Back firing may also be caused by the cylinder becoming overheated, or by projections, or fins, on the inside of the cylinder becoming incandescent and holding their heat, derived from one explosion, long enough to ignite the next partially compressed charge. A particle of carbonized oil may become incandescent with the same result. The spark may be too far advanced for the speed at which the engine is running.

In the two-cycle type of motor, premature explosions may occur in the crank case, due to its compression being poor, as has been previously explained.

Premature explosions are accompanied by pounding of the engine bearings, although a bearing pounding does not necessarily indicate that premature explosions are taking place, as a bolt or nut may be loose and produce the same result.

*Water in the Cylinder.*—This may result from water being introduced in the mixture, or, as is sometimes the case when the engine is made with a detachable cylinder head, from the gasket blowing from the cylinder into the water space. The condition is accompanied by loss of power, or, as is usually the case, by stopping of the engine. The igniter mechanism, if electrical, becomes grounded. The only remedy is to repack the head, an operation often done the wrong way by those inexperienced. The packing, which should be a good grade of wire-woven asbestos, should be carefully cut and fitted to the cylinder head, being sure to provide the openings for the water spaces and any others that may occur. Carefully cut all bolt holes, making them large enough to permit the bolt to pass through freely, without drawing the gasket out of place when they are screwed up. Cut all openings as nearly to the exact size as possible. Now place the gasket carefully in place on the head or cylinder, as is most convenient; place the head in position and insert the screws or bolts. With the engine *cold*, draw up the bolts as tight as they will go, of course using a reasonable amount of judgment and not twisting the heads off. Now, with the jacket dry, run the engine for 3 or 4 min., or until it is good and warm; this will soften the rubber, or other cementing material in the gasket, and allow the bolts or screws to be tightened up to their final position. Failure to perform this last tightening operation will mean that the gasket will blow again, as, when the engine is hot, it sometimes takes three-quarters of a turn on the screw, to take up the gasket and squeeze the cement into all the cracks so as to produce an absolutely tight joint.

*Engine Smokes.*—Smoke, issuing from the exhaust, indicates too rich a mixture or too much oil. Smoke, issuing from the front of the cylinder, indicates that the piston is leaking, due to the rings being worn or the cylinder out of round, or the engine may be running hot. The remedy for these conditions has already been mentioned.

*Valves Leak.*—See if the stems are sticking, or if the seat is crusted or cut, or if a spring is weak. The remedy has been previously suggested.

*Engine Races.*—If the engine, running light, races, or runs faster than it can be supplied with gas, it is an indication that the spark is too far advanced for the amount of mixture being fed to the cylinder. The remedy is to retard the spark or give the engine more mixture.

# CHAPTER VI.

## STARTING DEVICES.

THERE are a number of different methods in use for starting gas engines, all of which are used, more or less extensively, as requirements demand.

They may be enumerated about as follows: 1. Hand starting, which is used most extensively in the starting of engines of moderate size, and requires that the engine be provided with compression relief cocks or starting cams. In using this method of starting, care must be taken that the ignition is so set that the charge will not be prematurely exploded, causing back firing, with accompanying dangerous results to the operator. Engines which are to be started by this method are generally provided with an automatic throw-out collar which enables the operator to clutch the shaft with the starting crank, but which, when the engine starts, automatically throws the crank out of connection. Several devices, of greater or less efficiency, have been placed on the market, the object of which is to cause the starting crank to disengage as the engine starts or back fires, thus insuring immunity to the operator.

2. It is sometimes possible in multi-cylinder engines, and even at times in engines of but one cylinder, to start, after a moderately short stop, by retarding the sparking apparatus and igniting a cylinder containing part of a charge drawn in before the engine was stopped. To do this successfully demands that the engine be stopped with the spark and that the piston rings be a tight fit, insuring a tight cylinder.

3. The engine may be turned over until it takes up its cycle of operations by some external source of energy. Electric motors are often used to advantage for this purpose, or the large engine may be provided with a starting engine

small enough to be turned over by hand and of sufficient power, when running, to start the large engine.

4. An explosive mixture of gas and air may be stored in an auxiliary air-tight chamber. This may be accomplished by the engine itself charging this receptacle before it is stopped. On starting the engine, it is turned over dead center into the explosion stroke, and a charge of the explosive mixture is admitted to the cylinder by opening a valve in the supply pipe. The explosion of this charge will generally be sufficient to give the engine enough impulse to make it take up its cycle of operations.

5. The last-named method may be varied by using an air pump, operated by hand, to compress a charge of explosive mixture into the cylinder. In either of these last two methods named, the charge may be exploded by an electrical spark, if the electrical system of ignition be used, or by means of a match starter; see Fig. 6.

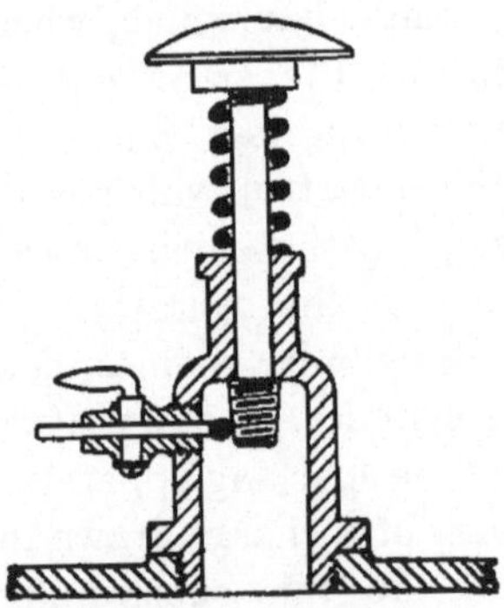

Fig. 6. Match Igniter.

6. The method of inserting an explosive cartridge in a tube, opening into the cylinder, and exploding it by mechanical means, has been used to some extent.

7. A charge may be exploded in an auxiliary chamber and the resulting pressure conveyed to the engine cylinder. Hutton, in his treatise on "The Gas Engine," illustrates such a starter, the operation of which is shown in Fig. 7. Gas enters the auxiliary chamber, *A*, through the supply pipe, *B*, and, the poppet valve, *C*, on the engine being closed, passes out through the cock, D, where it is ignited. As long as the gas valve in the supply pipe is kept open, the pressure in the explosion chamber is maintained sufficiently high to prevent the flame at the jet, D, from running back into it; but as soon as the supply is cut off, the gas in the chamber is gradually consumed at the jet until the

mixture is such that the flame runs back and ignites the entire charge, and the resulting pressure is admitted to the cylinder through the poppet valve, *C*.

8. By far the most widely used method of starting is by means of compressed air, compressed and stored by the engine itself or by means of a smaller auxiliary air compressor.

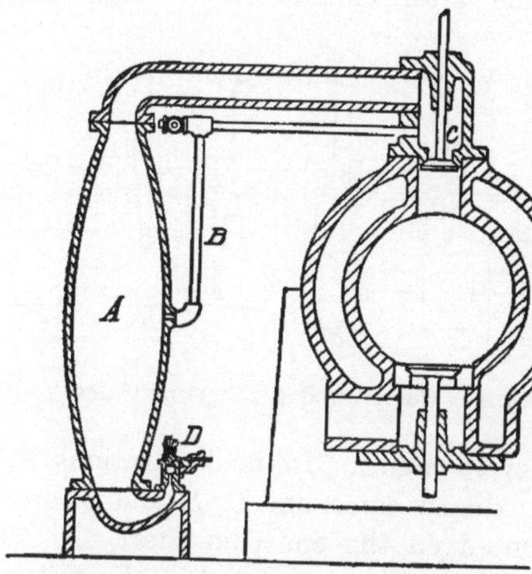


**Fig. 7. Auxiliary Chamber Starter.**

In order to operate the compressed-air starter, it is necessary that the cam movement in a four-cycle engine be so arranged that one or more of the cylinders may be converted, for starting purposes, into a compressed-air engine. To accomplish this it is necessary that the cylinder exhaust once every revolution and that the inlet valve remain continually closed while the starting operation is proceeding. The air may then be turned into the cylinder, either by means of an air cock actuated by hand or by automatic means. Where a single cylinder, in a multiple-cylinder engine, is thus arranged, it is necessary that the engine be turned over by hand until the piston of the air cylinder is at the beginning of the working stroke, when the starting mechanism may be thrown into gear and the air admitted.

As mentioned in Chapter XIV, if the cams are made to

operate an intermediate lever mounted on a shaft, which mechanism, in turn, operates the valve rod, it is an easy matter to make the shaft, on which the levers are mounted, to shift in such a way as to bring the starting mechanism into gear.

Such an arrangement is shown in Figs. 8 and 9, which figures show the arrangement of the cams for a three-

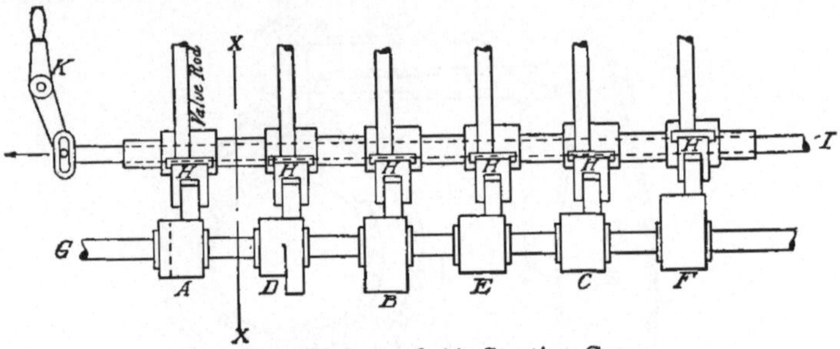


**Fig. 8. Compressed Air Starting Cams.**

cylinder, four-cycle engine. In the illustrations *A*, *B* and *C* are the three exhaust cams, and D, *E* and *F* are the inlet cams, all mounted on the one cam shaft, *G*. The transmission levers, *H*, are shown mounted on the shifter shaft, *I*. On the transmission levers are the hardened steel contacts,

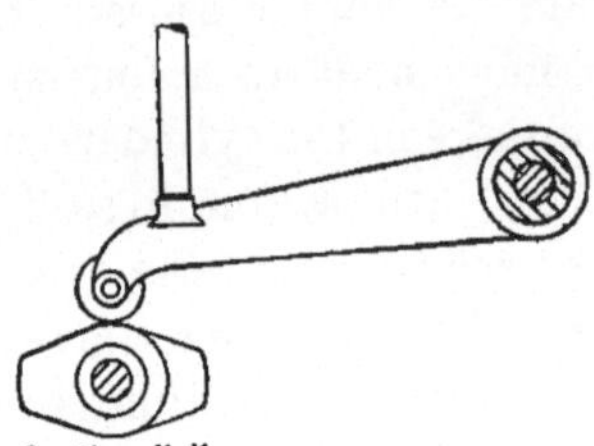


Fig. 9. Detail of Starting Cam shown in Fig. 8.

*J*, on which the valve stems impinge. Cams *A* and D are the double starting cams. Cam *A* is provided with two eccentric portions, as shown, so that the exhaust valve is made to open once every revolution, while cam D is made with one-half of it with the outline of the regular inlet cam, while the other half is concentric to the shaft at all points, so that when the lever, *H*, acts on the concentric portion it will not raise, and, in consequence, the inlet valve will remain closed. In starting, the engine is turned over until the cylinder, which the cams, *A* and *D*, control, is in the explosion stroke; then, by means of the lever, *K*, the shifter shaft carrying the levers, *H*, is moved in the direction of the arrow to the starting position and the air cock opened. A double cam, mounted on the cam shaft, is provided by some manufacturers to open and close an air poppet valve. This, however, is not an absolute necessity, although more economical of air, as it shuts off the air supply during the exhaust stroke of the piston, when the exhaust valve is open to the atmosphere.

With the air-starting system a pressure of from 200–300 lb. is maintained in the storage tank.

# CHAPTER VII.

## CARBURETTORS, VAPORIZERS AND INJECTORS.

The gas, gasoline, alcohol, and oil engines operate on the same general principle, as far as the generation of power is concerned, but the methods pursued for obtaining the requisite fuel in gaseous form vary with the several different types. Thus, in the gas engine, that is, the engines which operate on some form of gas as a fuel, no intermediate steps are necessary for the transformation of the fuel from a liquid to a vaporous or gaseous form, although, in the engines operating on producer gas, an apparatus, known as a producer, is necessary to distill from the fuel, as it appears in a solid state, a gas, available for use in internal-combustion engines.

The different devices used for the production of the combustible gasoline, alcohol, or oil mixture will first be discussed, after which the operation of a suction gas producer will be taken up.

There are three general methods in use for securing an explosive mixture from liquid fuels:

(1) Carburetting, (2) Vaporizing, (3) Injecting.

The carburettor, in any one of its many forms, is a device by which the liquid fuel is transformed into a vapor by passing air either over, through, or across a portion of the supply and taking up particles of the liquid in a vapor form. To facilitate the operation, when carburetting gasoline, it is much better, although not absolutely necessary, that the air, as well as the gasoline, be warm, especially in cold weather; and for this reason we find the engine manufacturers leading their suction from a "hot-box," located either on the exhaust manifold or on the cylinder base, or where the warm water, from the jacket, may be made to circulate around it.

When carburetting alcohol it is necessary, under all circumstances, that the fuel be warmed; the reason for this is fully explained in Chapter IX.

In carburetting petroleums it is necessary, especially with the heavier grades, that the air under pressure be forced through the liquid in order that it may break up or pulverize the fuel and carry a portion of it, in suspension, to the engine cylinder. In many engines, after the oil is thus broken up, the mixture is carried to a heated chamber or through heated coils where it is vaporized and mixed with air to form the proper explosive mixture. In other types, the vaporized fuel is carried direct to the cylinder, and the residual heat of previous explosions produces the same general results, although in a less satisfactory manner. Still another method consists in heating the fuel oil, by passing it through coils exposed to the action of the exhaust gases, and thereby driving off an oily vapor which, due to its heat, has sufficient pressure to carry it past an air nozzle where sufficient air is mixed with it to produce the proper explosive mixture. The same result is also obtained by causing the oil to fall, a drop at a time, on a hot plate, thus causing it to vaporize.

The vaporizer (and it may here be said that they are only applicable, as ordinarily designed, to the use of gasoline or naphtha) differs from the carburettor in that the latter always has a supply of gas on hand, while the vaporizer, or mixing valve, makes only enough gas for each revolution or charge, as required. Many, so-called, carburettors are, in reality, improved types of mixing valves, and in fact, it may be said that the majority of them are.

The vaporizer consists, essentially, of a gasoline valve, of needle design, capable of being adjusted to deliver the necessary amount of the fuel to produce the requisite vapor for the mixture required, and an adjustable air valve, by means of which the air supply may be regulated so as to vary the quality of the mixture, as requirements demand.

The gasoline is dropped in the path of the entering air and carried along in the form of a finely divided spray, or it is made to rise in a nozzle, placed in the path of the entering air, which carries enough of the fuel with it to produce the necessary mixture.

Vaporizers are made supplied with a throttling device and in all respects similar to the majority of the, so-called, carburettors on the market with the exception that they are not supplied with an automatic float feed device.

Injecting, as the name implies, consists in injecting into the cylinder, or a chamber adjacent thereto, a quantity of the fuel mixed with the requisite amount of air. This method of introducing the fuel into the cylinder is practiced quite largely by the different oil-engine manufacturers. The Hornsby-Akroid and the Meitz and Weiss, the injection and ignition of whose charges are later described under "Ignition," make use of this principle, as does also the Diesel motor.

The Diesel method is as follows. Referring to Fig. 10, the air valve, the exhaust valve, and the fuel valve are plainly marked. The air valve allows the air charge, as previously described, to enter the cylinder on the suction stroke. At the beginning of the working stroke the fuel valve is opened and the charge of oil is forced into the cylinder by means of compressed air under a pressure of 800 lb. per sq. in. The construction of the fuel valve is somewhat unique; the fuel enters the valve through the pipe, *A*, and the auxiliary compressed air, through the pipe, *B*. The valve, proper, consists of concentric washers, *C*, drilled with small holes, as shown, parallel to the spindle, D, which, by means of the governor acting through the bell crank, *E*, opens and closes the valve at *F*. The capillary attraction of the oil, as it falls on the washers, causes it to fill the holes above mentioned, and when, on the working stroke, the valve, *F*, is opened, the oil is carried with the air in a finely divided spray to the cylinder where, as described in the chapter on "Ignition," the heated air contained in the cylinder completely vaporizes and ignites it. The valve stem is made of nickel steel, as it has been found by experience that where the packing abrades the spindle, as it is moved back and forth, it soon becomes worn and requires to be replaced.

This method, as used on the Diesel motors, is very economical of fuel and could be applied to engines of lower compression with probably as satisfactory results.

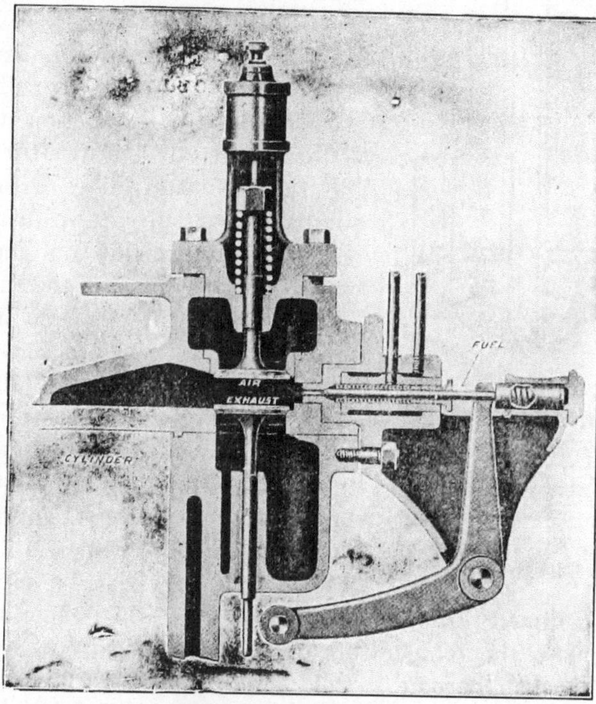


Fig. 10. Diesel Valves.

Returning to the subject of carburettors and referring to Fig. 11, we have an example in which the carburetted air is obtained by passing it through the fuel and thence to the engine. This method is sometimes spoken of as mechanical ebullition. In the illustration in question, *A* represents the suction pipe to the engine; *B*, the screened openings, through which the auxiliary air supply is drawn; *C*, the tube, terminating in the float, *E*, through which the carburetted air is drawn;

D, a shield acting on the principle of a separator and causing the surplus particles of gasoline to separate from the vapor when they impinge against its surface; *F*, an indicator and gauge by which the height of gasoline in the carburettor may be determined. The end of tube, *C*, terminates in such a position as always to be just below the surface of the liquid; the float, *E*, causing it to move up or down, as the elevation of the surface of the liquid is changed. On the suction stroke of the engine, air is drawn through the tube, *C*, and mixes with the auxiliary air supply drawn in at *B*, and the mixture thus obtained is carried to the cylinder of the engine.

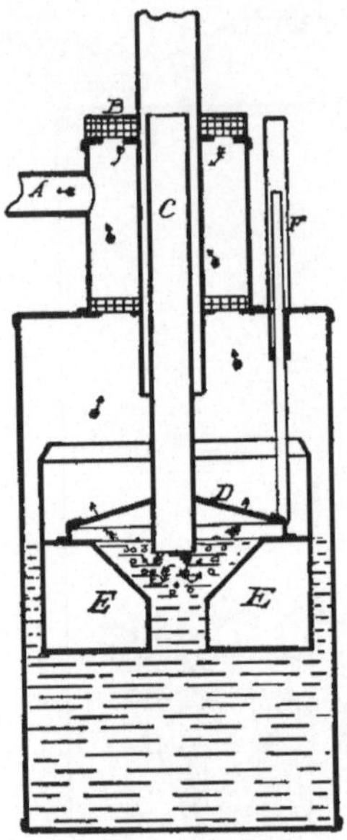

Fig. 11. Daimler Carburettor.

This carburettor was used in the earlier Daimler engines and was first devised by Gottlieb Daimler. There were, however, two very marked disadvantages in its use. In the vaporization of any liquid a certain amount of heat, known as the latent heat of vaporization, is lost; now unless heat be supplied to the liquid, from outside sources, as this vaporization continues, the temperature of the liquid continues to fall until it may become so cold that it will have lost nearly all of its volatility. (See Chapter IX for detailed description of this condition in different fluids.) This was the great difficulty encountered in the operation of this type of carburettor. Added to this, trouble was also experienced from fractional distillation; that is, the lighter portions of the liquid naturally rose to the top and, in the process of vaporization, passed off first, leaving in the carburettor the heavier part of the fuel. As a result of this condition the last part of the liquid to be vaporized, or distilled, was of poor quality, naturally affecting the operation of the engine.

Several carburettors of this general type were designed and used to some extent, but the same difficulties were experienced in all.

Fig. 12 represents one of the earlier carburettors, in which the carburetted charge was obtained by passing air across

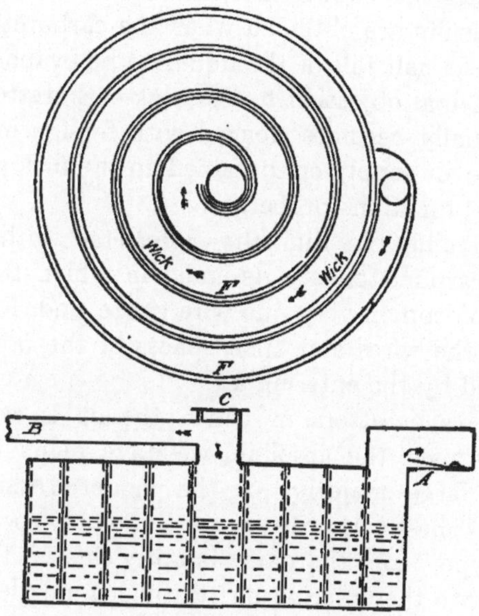


Fig. 12. Surface Carburettor.

a large surface on which the fuel to be vaporized was maintained.

In the illustration, *A* is the inlet valve, through which the air to be carburetted is drawn; *B* is the suction pipe to the engine; *C* is the auxiliary air supply valve, by means of which the mixture may be regulated; *E* is a light metal casing, containing the spiral *F*, which is fastened to the top of the casing and allowed to project nearly to the bottom, as shown. On either side of the spiral partition is basted light flannel or felt.

When the carburettor is partially filled with fuel, as shown, it is apparent that the air, entering at *A*, must pass completely through the spiral to reach the outlet at *B*. The flannel or felt, reaching down into the liquid, serves in the capacity of a wick, and, by capillary attraction, the part above the liquid is kept saturated with the fuel. The passage of the air currents across this large expanse of wick surface evaporates the fuel and saturates the air.

Experience has shown that, in carburettors of this type, the best results are obtained when the carburettor, if about 8 in. deep, is half full of the liquid to be evaporated.

The principal objection to the wick carburettor is that the wick gradually becomes clogged with foreign matter, taken in with the air supply or contained in the fuel, and ceases to perform its functions properly.

To overcome the difficulty connected with the use of wicking, carburettors are designed in which the liquid fuel is made to drop on very fine wire gauze, and, forming a thin film over the wires and open spaces of the fabric, is easily evaporated by the entering air.

Of the carburettors, in which the air is carburetted by passing it over the gasoline, we have many examples; in fact, the large majority of the modern designs use this method. These carburettors are commonly spoken of as the "spray type," and the name is aptly chosen, the fuel being injected into the entering air, through a nozzle, in the form of a finely divided spray or mist.

The operation of the carburettor is as follows: a nozzle, fitted with a needle valve, is so situated in the air passage that the entering air must pass over it on the way to the engine. On the suction stroke, a partial vacuum is formed in the carburettor, and the air, rushing over the spraying nozzle, is attempting to fill this vacuum. In consequence its pressure is less than that of the atmosphere, and the fuel, supported in the nozzle, is thrown off into it by the unbalanced pressure.

As has been previously noted, these carburettors have a capacity of but little more than one complete charge for the cylinder, and are really improved types of mixing valves, their one distinctive feature being the float feed, with which they are almost universally equipped. If not equipped with a float feed, an aspirating valve is necessary to automatically open and close the spraying nozzle.

Their greatest difficulty arises from the tendency of small particles of foreign matter to lodge in the spraying nozzle, and this fact necessitates great care in the filling of the fuel supply tank.

In the carburettors in which an automatic aspirating valve is used in place of the float feed, some difficulty is encountered, especially on high speeds, from the inertia of

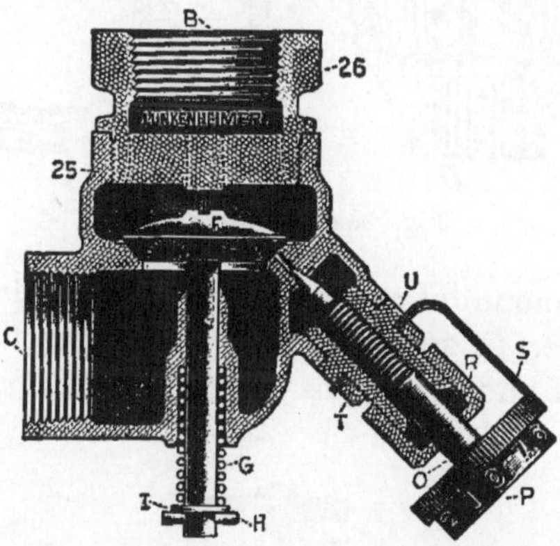


Fig. 13A. James-Lunkenheimer Mixing Valve.

the valve spring not allowing sufficient time, on every suction stroke, for the entering air to secure its requisite amount of vaporous fuel. As commonly spoken of, the carburettor is not flexible enough, and, on high speeds, the power of the engine is not proportional to the increased number of impulses received. For this reason the ordinary mixing valve gives, as a rule, a good deal of difficulty when used on a two-cycle engine of the three-port type.

The generic patent, under which nearly all the modern carburettors are manufactured, is shown in Fig. 13, and is known as the James-Lunkenheimer design. It consists essentially of a poppet valve, *A*, the movement of which may be restricted by means of the screw, *B*, and which is held to its seat by the spring, *C*. The air supply enters through the port, *D*, and the gasoline through the nozzle in the valve seat at $E$. The gasoline supply is regulated by means of the needle valve, the handle of which is shown at

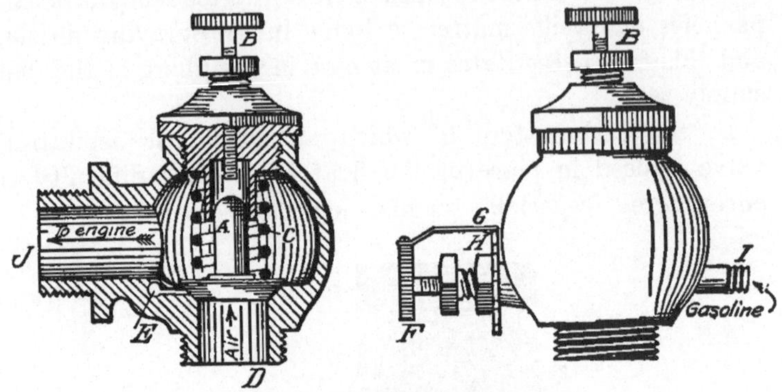


Fig. 13B. Mixing Valve.

$F$, and the amount of opening of the needle valve is indicated by the pointer, $G$, and the dial, $H$. The gasoline connection is shown at $I$ and the engine connection at $J$.

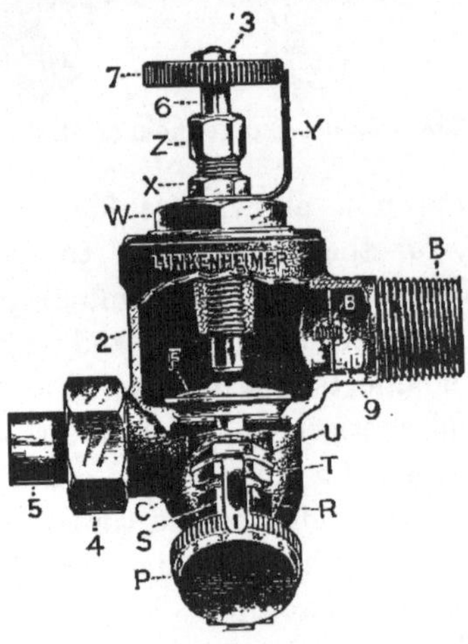


Fig. 14. Plain Pattern Generator Valve.

Figs. 14, 15 and 16 represent three different types of mixing valves manufactured by the Lunkenheimer Company, of Cincinnati, Ohio. Fig. 14 is the plain pattern generator valve, Fig. 15 shows a valve with screws for varying the opening of the poppet and the tension of the spring, and

**Fig. 15. Generator Valve with Adjustable Air Poppet.**

**Fig. 16. Generator Valve with Throttle.**

Fig. 16 is a valve having throttle connections, means for varying the air and gasoline supply and for changing the tension of the valve spring.

These valves, especially the one with throttle connection, give excellent satisfaction, and the operator is able to control the motor very closely.

With nearly all types of mixing valves the best results are obtained when they are operated on ordinary stove gasoline. This is due to the fact that the lack of the float feed allows the fuel to run into the valve more freely than is necessary, and, in consequence, the mixture is liable to become too rich with the more highly volatile gasoline.

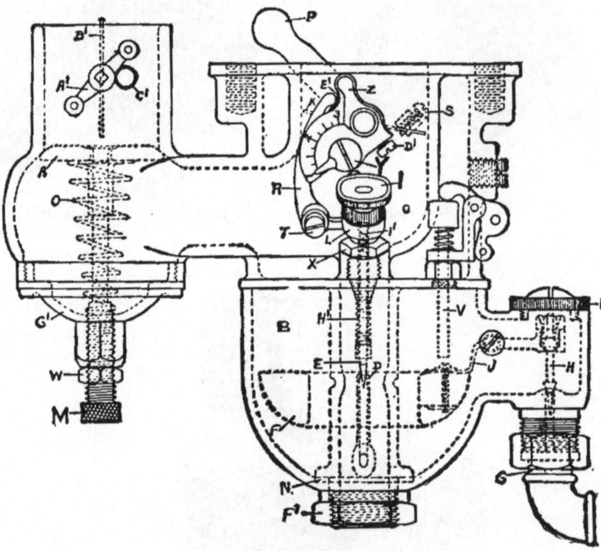

Fig. 17. The Schebler Carburettor.

Fig. 17 is a cut of the Schebler carburettor, manufactured by Wheeler & Schebler, Indianapolis, Indiana. It is of the spray-float feed type, and the working parts are shown quite clearly in the cut. Reference figures and letters are as follows: 9 is the constant air opening, through which the air, to be carburetted, passes. The air, entering at 9, passes upwards past the spraying nozzle *P*, to the mixing chamber of the carburettor, where the auxiliary air supply, entering through the poppet valve, *A*, is mixed with it, and the explosive mixture, thus formed, is drawn into the engine cylinder. The float feed mechanism is shown at *F*, *J*, *H*, and consists of a float, *F*, surrounding the constant air supply tube, as shown, which operates the needle valve, *H*, through the lever, *J*, maintaining the level of the fluid such that it will just overflow the nozzle, *P*. Gasoline enters the carburettor, from the supply tank, through the supply pipe, *G*.

The nozzle, *P*, is fitted with a needle valve, *E*, which is adjusted permanently for low throttle by means of the knurled button, *I*; then for open throttle, the needle valve mechanism is raised or lowered, by the operator, by means of the lever, *P*, which actuates the cam mechanism, *Y*, and causes the lever, *Q*, to revolve about the point, *T*, thus raising or lowering the needle point. The auxiliary air supply is provided with two valves, a damper valve *Z*1, by means of which the opening may be increased or diminished; and the poppet valve *A*, held to its seat by the tension of spring *O*, which tension may be increased or diminished by means of the knurled screw *M*. Push pin *U* is used for priming the carburettor when starting the engine. Pushing the pin down lowers the float and opens the needle valve, *H*, causing the nozzle, *P*, to overflow and producing, momentarily, a very rich mixture, suitable for starting. The float chamber is usually provided with a pet-cock, at its lowest point, for draining off the poorer grade of gasoline, which usually accumulates there. If the gasoline contains water it also accumulates in the bottom of the float chamber, and may be removed from time to time.

Fig. 18 shows two sectional views of the Holley carburettor, manufactured by the Holley Brothers Company, Detroit, Michigan. It is also of the spray type, and its operation is as follows: the incoming air enters through port *A*, which is provided with a fiber valve *BC*; situated around the valve are the constant openings *a*, through which air is constantly passed. The valve proper is held to its seat by means of the spring, *b*, as shown, the tension of the spring being capable of adjustment by means of the adjusting screw *c*. The spraying nozzle, D, situated in the path of the incoming air, may be opened or closed by means of the needle valve, *d*. The mixture passes to the engine past the butterfly throttle, *E*, which may be opened or closed, to suit requirements, by means of the lever, *F*. Gasoline enters the carburettor through the gasoline connection, *H*, which is opened or closed by means of the needle valve, *I*, being raised or

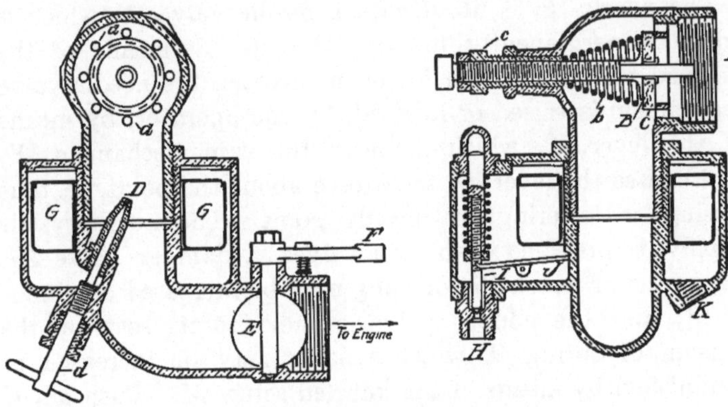


Fig. 18. Old Type Holley Carburettor.

lowered by the float, *G*, acting through the lever, *J*. The mechanism is set to maintain the height of the liquid such that it will just overflow the spraying nozzle. On low speeds the air enters entirely through the constant air openings, *a*, but, as the speed increases, these openings not being large enough to supply the requisite amount of air to overcome the partial vacuum formed, the auxiliary valve *BC* raises and allows an extra supply to enter the carburettor. A pet-cock at *K* is used to drain the carburettor, as already described.

Fig. 19 is a later design of the Holley carburettor in which an entirely new principle has been made use of to produce the varying mixtures necessary for the changing speeds. More or less trouble is encountered with carburettors, in which this variation is accomplished by means of an auxiliary air supply, owing to derangements of the spring device, which closes the auxiliary valve. Frequent adjustments of this

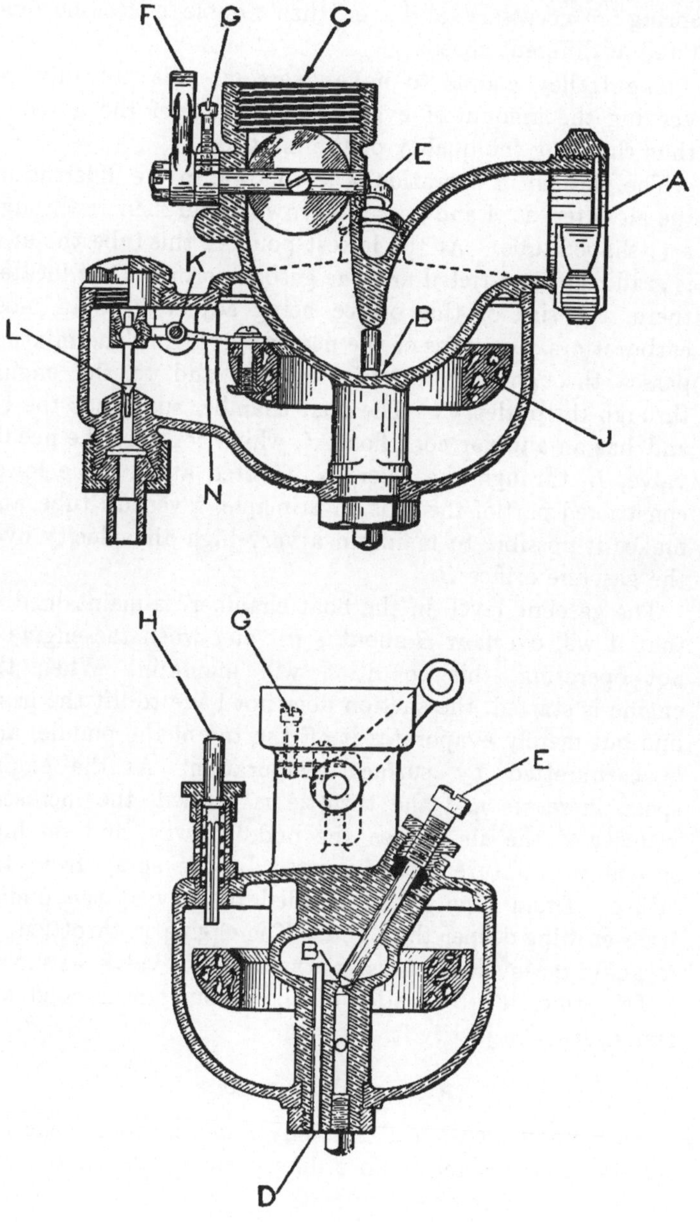


Fig. 19. Holley Carburettor.

spring are necessary, and even then trouble is often encountered at different speeds.

The Holley claims to have overcome this difficulty by varying the amount of evaporating surface of the gasoline, thus changing the quality of the mixture.

The method of operation is as follows: in the illustration, the air enters at *A* and passes downward and then up through a U-shaped tube. At the lowest point of this tube the area is gradually constricted and the gasoline orifice, *B*, is located there, the size of this orifice being adjusted, as in most carburettors, by means of the needle valve, *E*. The mixture passes through the butterfly throttle and to the engine through the outlet, *C*. The float chamber surrounds the U, and has an annular cork float, *J*, which controls the needle valve, *L*, through the lever, *N*, pivoted at *K*. The lower constricted part of the U is, in principle, a venturi tube, and makes it possible to maintain a very high air velocity over the gasoline orifice *B*.

The gasoline level, in the float chamber, is maintained so that it will overflow *B* about $\frac{1}{8}$ in., and when the engine is not operating, this condition will maintain. When the engine is started, the suction does not have to lift the gasoline but merely evaporates it off the top of the puddle, and is carburetted by surface evaporation. As the engine speed increases and the throttle is opened, the increased velocity of the air sweeps the puddle away, and on high speeds the mixture is carburetted by the spray from the orifice. Drain pipe D is provided to prevent the puddle from growing deeper than $\frac{1}{8}$ in. The engine is throttled by means of the lever, *F*, operating the butterfly valve, as shown.

The principle of operation of this carburettor is good, and results seem to justify it.

## Alcohol Carburettors.

As already mentioned, the only requirement for the carburetting of alcohol in an ordinary carburettor is that the air shall be of sufficient warmth to vaporize the alcohol in sufficient quantities to produce a properly saturated mixture.

However, it has been found in many cases to be the fact that the suction should be increased when carburetting alcohol, owing to the increased amount of fuel used to produce the same power.

In Germany and France double carburettors are in use to some extent. The engine is started on gasoline and, when sufficient heat has been generated to insure the perfect operation of the alcohol, a valve cuts out the gasoline supply and cuts in the alcohol.

Other carburettors for use with alcohol utilize the exhaust gases to heat and vaporize the fuel and when in a vaporous state it is mixed with the air and passes to the engine.

It is apparent that carburettors, through which the exhaust gases must pass, are more or less cumbersome and it is doubtful if they would be as efficient, as producers of power at the engine cylinder, as those in which the fuel is not heated to as high a temperature. It was proven conclusively, by tests made on gasoline engines, that this was so in their case, and there is no good reason why the same rule should not hold true in the use of alcohol, for the hotter the entering charge, the less the range of temperature in the cylinder and the less the power developed. Moreover, since government tests have seemed to show that most gasoline carburettors are, with slight alterations, adaptable to the use of alcohol there is no valid reason for making them more complicated.

### Carburettor Design.

A jet carburettor, in order to give satisfaction, should have an air velocity at the jet sufficient to secure a good spraying effect. The vacuum in a carburettor operating on alcohol, should be, according to best authority, equal to between $\frac{3}{4}$ in. and $\frac{7}{8}$ in. of mercury, and for starting, this value should be about doubled. In a carburettor operating on gasoline, a vacuum of 0.1 in. of mercury is generally sufficient.

The speed of the gases passing the spraying nozzle should be, in the use of gasoline, about 80 ft. per sec., while to satisfactorily spray alcohol into the passing air the velocity should be about 220 ft. per sec.

## CHAPTER VIII.

### PRODUCERS.

PRODUCER gas fuel, for gas engines, may be generated with apparatus operating under pressure, or by suction.

The first producers to be made and marketed were introduced by a Londoner, named Dowson, and they were used to such an extent that the name Dowson gas came to be almost synonymous with producer gas. They were of the pressure type and required for their operation a hard grade of anthracite. The producer plants were quite complicated, due to the necessity of scrubbers, cooling apparatus, and a gasometer, in which the gas, since it was under somewhat varying pressure, had to be stored before being fed to the engine.

There are many cases, however, in which it is necessary that a pressure system be used; in fact for any other purpose than for use in connection with a gas engine, where the suction of the piston produces the necessary flow of air through the producer, pressure of air is necessary in order to operate the apparatus and to convey the gas to the required point.

Nearly all pressure generators are copies of the original Dowson idea and include a generator or retort, in which the gas is driven off from the fuel; an air-compressor fan or other apparatus for blowing a mixture of steam and air through the generator; a scrubber, a gas purifier, and a gasometer.

Fig. 20 represents, diagrammatically, such a plant.

The retort or producer consists of a metal shell, lined with fire brick or clay, vertically mounted. A charging hopper, so arranged that the producer is never opened to the air, is provided at the top. The bottom of the producer rests on a grate through which the ashes fall, the air and steam being passed through these ashes to the producer, or a water seal is provided, in which the generator sits, and the mixture of air and steam is introduced under a conical hood which protects the open end of the pipe from becoming clogged

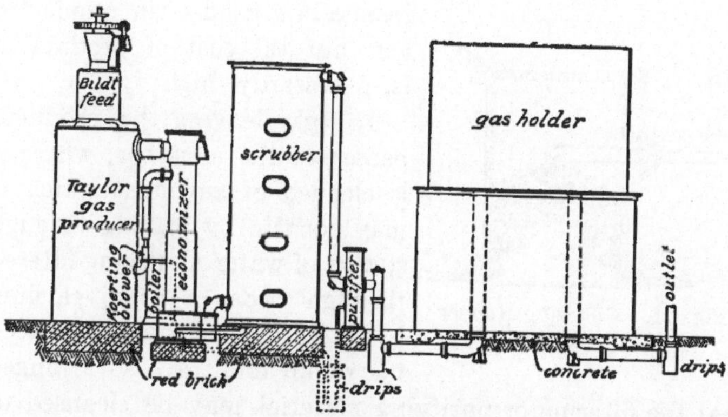


Fig. 20. Pressure Producer.

with ashes and coal. The mixture should be superheated if possible, in order that no more heat energy than is necessary be lost in heating the entering air.

The entering fuel should be carefully distributed, and means provided for breaking the clinker formation on the walls of the retort; some manufacturers provide tuyere openings to accomplish this; the apparatus must be tight and, if not provided with a water seal, suitable means must be had for cleaning the grate.

Distribution of the fuel charge is accomplished by making the drop grate of the hopper conical in form, which spreads the coal over a large surface. (See Fig. 21.)

The air blast may be supplied to the producer in any one of several different ways. The pressure may be obtained direct from a steam boiler maintained, as nearly as possible, at 80 lb. per sq. in. pressure; a blower, operating on the principle of a draft inducer or injector; by means of a mechanical fan or centrifugal blower; or by the use of compressed air.

When blowing with steam, trouble is more than likely to be encountered from the varying steam pressure producing different qualities of gas. Mechanical blowers may be driven directly from the engine, if used in connection with a producer for a gas engine. The use of compressed air results in a good even quality of gas, but the cost of production is, necessarily, high.

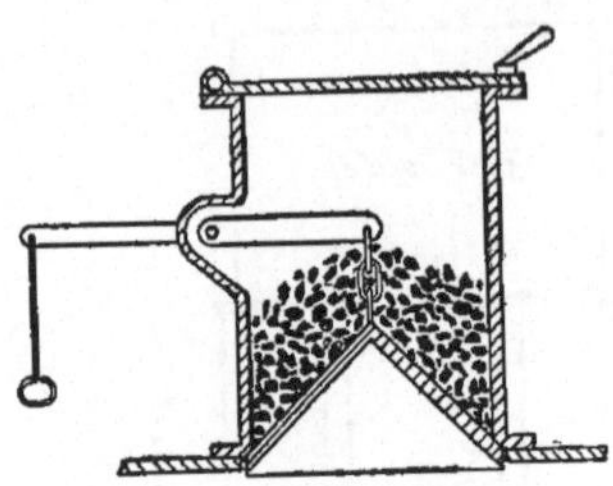

Fig. 21. Charging Hopper for Producer.

The gas leaving the producer passes to the scrubber, where it is cleaned of any dust which it may contain by passing through sprays of water and being filtered through beds of coke, calcium hydrate, moss, or sawdust, placed on removable trays so arranged that the filtering or purifying material may be cleansed or renewed.

From the scrubber and purifier the gas passes on to the gasometer, where it is stored ready for use. The gasometer acts in the capacity of a pressure regulator and should have sufficient capacity to take care of any possible stopping of the production of gas for a brief time.

In producers operating on the pressure system, any combustible or volatile material may be used to produce the gaseous fuel, and in nearly all cases the economy over the combustion of the same material under a steam boiler is very marked. This gas may be produced from sawdust, sawmill refuse, street sweepings, garbage, lignite, peat, etc. The methods used for producing the gas vary to some extent — the gas being obtained either by distillation or combustion. Since these waste products and the cheaper grades of coal or peat in burning form a rather closely compacted mass, considerable pressure is required to drive the air through the producer; hence, without exception, these materials are not available for suction producers.

In the production of gas from wood or wood refuse, by the distillation process, the material is placed in a cast-iron crucible which is subjected to the heat of a furnace and the volatile part of the fuel distilled off, leaving charcoal in the crucible. The walls of the crucible should be heated to a cherry-red heat, between 1600 degrees and 1700 degrees fahr., and its diameter should not exceed 12 in.

Fig. 22 is a cut of the Riche distilling producer as illustrated by Mathot in his "Modern Gas Engines and Producer Gas Plants." The heated gases from the furnace pass through the flue opening, *A*, into the flue space, *B*, which surrounds the retort, as shown; the gases then circulate around the retort and pass up and out to the stack through port, *C*. The crucible is charged with the fuel to be distilled, in this case pieces of wood, and the top closed to make an air-tight joint. From the bottom of the crucible the gas generated is led to the scrubber and purifier and then to the gasometer. The heat generated in the crucible raises the pressure of the products being distilled and, since there is no outlet at the top, the gases must pass from the cooler part past the hot part of the apparatus on the way to the scrubber. This has a tendency to burn out the impurities contained.

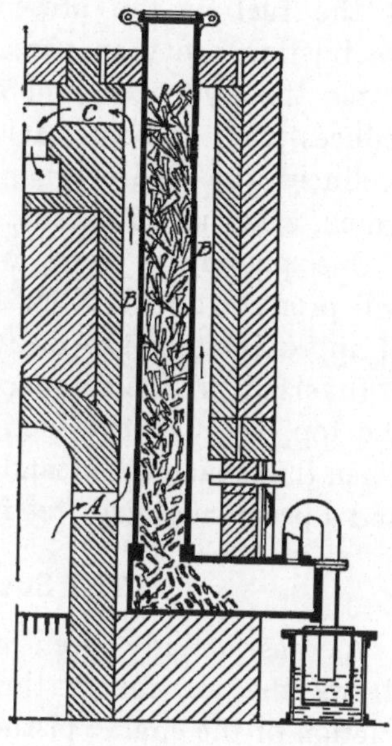

Fig. 22. The Riche Distilling Producer.

Producers operating on the distillation principle burn about 1 lb. of coal to every 2.5 to 3 lb. of material distilled, and produce from 28 to 35 cu. ft. of gas having a heating value of about 340 B.t.u. per cu. ft. or 9860 B.t.u. per pound of coal, whereas 1 lb. of good coal will produce of itself about 10,000 B.t.u.

The gas produced from the wood, however, is of permanent composition and can be transported long distances. The residual charcoal, if wood be used, also has some value. The residual weight of charcoal is approximately one-fifth the original weight of the wood — depending, however, to quite an extent on the amount of water in the wood, a wood like elm containing a very large percentage.

Combustion producers produce the gas by the combustion of the fuel in the presence of water. The products of combustion are then passed on to a reducer, which dissociates the hydrogen and oxygen contained in the steam, reduces the carbonic acid gas to carbon monoxide, and produces the hydrocarbons. The reducer contains coke, which, when incandescent, produces the necessary reactions.

One pound of wood waste, in a combustion reducer, will produce about 10 cu. ft. of gas having a heating value of approximately 115 B.t.u. per cu. ft.

Inverted producers operate by forcing the air down from the top, through the fuel. The distilled volatile products, when they reach the incandescent part of the fuel, are reduced and a permanent gas, free from tar, is obtained.

### The Suction Producer.

As has already been mentioned, the suction producer draws the air charge through the fuel by means of the suction of the engine piston and, in consequence of this fact, only certain fuels are available for use in these plants. The pressure type, it is readily seen, has greater elasticity in meeting the different fuel conditions, but the suction plant takes up much less floor space and the cost of installation is much less than for a pressure plant of the same size.

The suction producer can use only anthracite coal, or carbonized fuels, as charcoal or coke. The anthracite must not be too small — not less than pea size — and it must be clean and carry as small a percentage of ash as possible, not more than 15 per cent. Undue resistance in the producer will produce an over amount of back suction on the engine piston, with consequent loss of power.

If anthracite coal is used, one pound of good quality will produce about 1 b. hp.-hr., and it will require about 1.3 lb. of coke to produce the same result. For small units the amount of fuel is somewhat increased and may be as high as 1.5 lb. per b. hp.-hr.

Fig. 23. Suction Producer.

In operating the suction producer, a fire is kindled on the grate and a bed is built upon it in the ordinary manner, the necessary air being supplied, for the time being, with a hand or belt-driven fan. Beyond the vaporizer, the products of combustion escape through a waste pipe. When the test cock shows that a good quality of gas is being produced, the scrubber and purifier is thrown into the circuit, and when good gas appears at the engine cock, the engine is started and the fan stopped.

Fig. 23 illustrates a suction producer manufactured by R. D. Wood & Co., of Philadelphia. It is very compact, an area of 15 by 35 ft. being sufficient for a plant of several hundred horsepower, a smaller plant, of course, requiring less space.

Fig. 24 shows very clearly the comparative efficiencies of a gas producer and steam plant.*

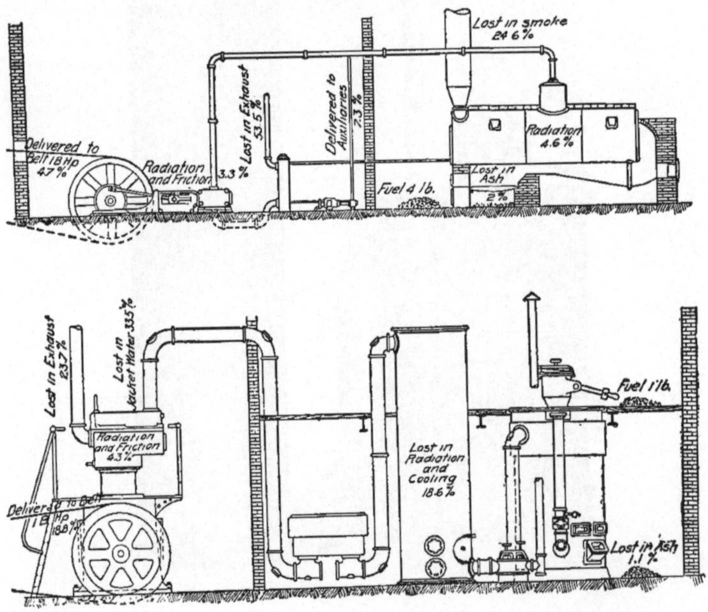


Fig. 24. Comparative Efficiencies of Steam and Producer Plants.

The first cost of a producer plant is approximately the same as for first-class steam engines and boilers of the same horsepower, but the resultant economies in fuel and attention are very marked, one man being able to care for a large plant. The cost of attention is from 50 to 75 per cent that of a steam plant. No time is required, as with the steam

* The author is indebted to R. D. Wood & Co. for the cuts on suction producers, and the results shown in their diagram are very closely in keeping with the results of modern practice where the steam engine used is of non-expansion, non-condensing type.

plant, to get up a head of steam, and when the engine is stopped for any length of time the producer may also be shut down.

The gas obtained from a suction plant operating with good fuel has a heat value of about 145 B.t.u. per cu. ft. The following analysis gives the approximate composition:

| | |
|---|---|
| Carbon dioxide, $CO_2$ | 0.7 to 0.9 |
| Carbon monoxide, CO | 0.24 to 0.28 |
| Hydrogen, H | 0.16 to 0.20 |
| Marsh gas, $CH_4$ | 0.04 to 0.06 |
| Nitrogen, N | 0.45 to 0.49 |

For large plants or plants operating several engines the pressure producer is advisable, but for use in connection with a single unit of moderate size the suction plant is cheaper, occupies less floor space, and serves the purpose equally as well.

# CHAPTER IX.

## FUELS AND COMBUSTION.

For the motive power in an internal-combustion engine any gaseous fuel is available, as well as any other fuel which may be vaporized or transformed into a gas. By vaporized fuel we mean any fuel, such as gasoline, petroleum, oil, or alcohol, which may be used in the cylinder of a gas engine without the intermediate step of transforming it into a gas. There is absolutely no combustible substance which may not be transformed into a gas, or ràther have its gaseous products driven off, by the action of heat. Any one of these gases may be used, with greater or less efficiency, as their calorific efficiency is greater or less, in the cylinder of a gas engine. It is furthermore true that in all cases the power obtained from any fuel first converted into a gas and then burned in the cylinder of an internal-combustion engine, is always greater than if the same amount of fuel were burned under a boiler and the steam used to drive a steam engine. This is true because the heat efficiency of a gas engine is about 25 per cent, while that of the steam engine is from 10 to 12 per cent. As nearly all of the combustible part of the fuel becomes gas — especially is this true in the case of vaporized fuels — it is obvious that the fuel so used must be much more economical than when fired under a boiler. The manufacturers' guarantee accompanying stationary gas engines usually insures that their engine will produce 1 hp.-hr. on from 11 to 12 cu. ft. of natural gas. The heat units contained in natural gas range from 900 to 1100 per cu. ft., according to the nature of the gas, and in producer gas there are about 160 heat units per cu. ft. Running on producer gas, then, the same engine would require

$11 \times \frac{1100}{160} = 75$ cu. ft. approximately. Now one pound of bituminous coal will produce 75 cu. ft. of gas and, in consequence, to produce 1 hp.-hr. would require that 1 lb. of coal be used in the producer. The very best steam engines yield but 1 hp.-hr. on 1.5 lb. of coal.

If we could use the relative heat efficiencies of a gas and a steam engine as a basis of comparison, we would expect to find that where 1 lb. of coal, as gas, in a gas-engine cylinder, would produce 1 hp.-hr., it would require 2 lb. of coal, or a trifle more, when used under a steam boiler. This comparison may not be used in the comparison of producer gas, as a certain amount of the available heat is used in the producer in distilling off the gas and in vaporizing the water contained in the fuel. As regards plant construction and operation, the gas producer is much cheaper and simpler an apparatus than the steam boiler and requires less attention. A surplus of gas may always be kept on hand, in a gasometer or storage tank, and the engine started on a moment's notice, while with a steam boiler time must be consumed in getting up a head of steam.

Gas too lean to be used under a boiler is found to ignite rapidly when under compression in a gas-engine cylinder. Blast-furnace gas is an example of the above, large two-cycle engines operating on this gas being in use at the shops of the Lackawanna Steel Company, in Buffalo, New York.

The value of a gas as used in a gas-engine cylinder is largely dependent on the number of British thermal units it contains, although the richness of the gas must also be considered. A lean gas may be burned completely in a cylinder with less air, and a consequent larger amount of gas, than a gas of high thermal value. The following table gives the heat units per pound and per cubic foot for the different fuels. Natural gas is seen to have the greatest heat value, but, notwithstanding this, gasoline vapor, with its lower heat value, is credited with about 11 per cent more power. This fact is due to the rate of flame propagation being more rapid in the gasoline vapor than in the use of natural gas, with the result that the combustion assumes more the aspect of an explosion. A corresponding increase in the mean effective pressure results.

TABLE I.

HEATING VALUES OF DIFFERENT FUELS.

| Fuel. | Molecular Weight. | Heat Units per Lb. | Heat Units per Cu. Ft. |
|---|---|---|---|
| Hydrogen, H | 1 | 61,560 | 293.5 |
| Carbon, C | 12 | 14,540 | ........ |
| Kerosene, $C_{10}H_{22}$ | 142 | 22,000 | ........ |
| Benzene, $C_6H_6$ | 78 | 18,448–22,000 | ........ |
| Gasoline, $C_6H_{14}$ | 86 | 18,000–21,000 | ........ |
| Alcohol, ethyl, $C_2H_5OH$ | 46 | 12,950 | ........ |
| Alcohol, methyl, $C_2H_4O_2$ | 60 | 20,000 | ........ |
| Acetylene, $C_2H_2$ | 26 | 21,492 | 868 |
| Gasoline vapor, $C_2H_{14}$ | 38 | 18,000 | ........ |
| Crude petroleum | ........ | 18,350 | ........ |
| Natural gas | ........ | ........ | 900–1,050 |
| Illuminating gas 19 c.p. | ........ | ........ | 800 |
| " 16 " | ........ | ........ | 665 |
| " 15 " | ........ | ........ | 620 |
| Water gas | ........ | ........ | 300 |
| Producer gas | ........ | ........ | 100–160 |
| Suction gas | ........ | ........ | 135 |
| Marsh gas (methane), $CH_4$ | 13 | 23,594 | 1,051 |
| Olefiant gas, ethylene, $C_2H_4$ | 28 | 21,430 | 1,677 |

The constituent parts of natural gases vary in different localities. Table II gives some of the different volumetric analyses of Pennsylvania gases.

TABLE II.

VOLUMETRIC ANALYSIS OF PENNSYLVANIA GASES.

| Constituents. | Burn's Butler Co. Well. | Pittsburg. | Harvey's Butler Co. Well. | Leech-burg. |
|---|---|---|---|---|
| Hydrogen, H | 6.10 | 22.00 | 13.50 | 4.79 |
| Ethylene, $C_2H_4$ | 18.12 | 5.00 | 5.72 | 4.39 |
| Marsh gas, $CH_4$ | 75.44 | 67.00 | 80.11 | 89.65 |
| Hydrocarbons (unnamed) | ........ | 1.00 | ........ | 0.56 |
| Carbonic acid, $CO_2$ | 0.34 | .60 | .66 | 0.35 |
| Carbonic oxide, CO | Trace | 0.60 | Trace | 0.26 |
| Oxygen | ........ | 0.80 | ........ | ........ |
| Nitrogen | ........ | 3.00 | ........ | ........ |
| Heat value, cu. ft. | 1,151 | 892 | 959 | 1,051 |

Many natural gases, especially those from western fields, have much lower heat values than those here tabulated.

*Oil Gas.* — Oil gas, as manufactured for municipal lighting plants, is of the following approximate composition:

| | |
|---|---|
| Hydrogen, H | 39.5 |
| Marsh-gas, $CH_4$ | 37.3 |
| Nitrogen, N | 8.2 |
| Hydrocarbons (unnamed) | 6.6 |
| Carbonic oxide, CO | 4.3 |
| Oxygen, O | 1.4 |
| Water vapor, $H_2O$ | 2.7 |

The heat value of this gas is slightly more than 600 heat units per cu. ft.

*Producer Gas.* — The so-called producer gases are of different volumetric compositions, as their methods of production vary. The following may be taken as average values:

| | Anthracite. | Bituminous. |
|---|---|---|
| Hydrogen, H | 12.0 | 12.0 |
| Marsh gas, $CH_4$ | 1.2 | 2.5 |
| Nitrogen, N | 57.0 | 55.3 |
| Carbonic oxide, CO | 27.0 | 27.0 |
| Ethylene, $C_2H_4$ | ........ | 0.4 |
| Carbonic acid, $CO_2$ | 2.5 | 2.5 |
| Oxygen, O | 0.3 | 0.3 |
| Heat value, cu. ft. | 137 | 156 |

The presence of the hydrocarbon element to a greater extent in the bituminous producer gas makes its heat value correspondingly greater.

*Water Gas.* — The production of this gas entails the loss of a large amount of energy and for this reason it does not play a very important part in the gas-engine field, although it is a gas very rich in energy itself. It is made by the action of a jet of steam on incandescent fuel, the hydrogen in the steam being dissociated and taken up by the gas.

R. D. Wood & Co. give the following data on the production of water gas, the theoretical composition of which is equal volumes of hydrogen and carbon monoxide, although the gas generally contains some nitrogen and carbonic acid:

| | Pounds. |
|---|---|
| 500 cu. ft. of H weigh | 2.635 |
| 500 cu. ft. of CO weigh | 36.89 |
| Total weight of 1,000 cu. ft. | 39.525 |

Now, as carbon monoxide is composed of 12 parts carbon to 16 of oxygen, the weight of carbon in 36.89 lb. of gas is 15.81 lb. and of oxygen 21.08 lb. When this oxygen is derived from water (steam) it liberates, as above, 26.35 lb. of hydrogen. The heat developed and absorbed in these reactions (disregarding the energy required to elevate the coal from the temperature of the atmosphere to say 1800 degrees) is as follows:

| | | Heat units. |
|---|---|---|
| 2.635 lb. of H absorb, in dissociation from water, 2.635 × 62,000 | = | 163,370 |
| 15.81 lb. C burned to CO develop 15.81 × 4,400 | = | 69,564 |
| Excess of heat absorption over heat development | = | 93,806 |

The loss due to this absorption must be made up in some way or other.

6.47 lb. of carbon burned to carbon dioxide would supply this heat, theoretically, but in practice, owing to the imperfect and indirect combustion and radiation, more than double this amount is required. Besides this it is not often that the sum of the carbon monoxide and hydrogen exceeds 90 per cent, the remainder being carbon dioxide and nitrogen.

The following is the average volumetric analysis of water gas:

| | |
|---|---|
| Carbonic oxide, CO | .41 |
| Hydrogen, H | .48 |
| Carbonic acid, $CO_2$ | .6 |
| Nitrogen, N | .5 |
| Heat value, cu. ft. | 291 |

Suction producer plants are quite largely in use in connection with the operation of gas engines. They are applicable to the use of anthracite only, depending for their operation on the suction of the engine instead of pressure on the producer. The gas thus produced is of good quality, with a heat value higher than pressure producer gas with anthracite fuel. Following is the analysis:

| | |
|---|---|
| Carbon dioxide, $CO_2$ | 8.0 |
| Carbon monoxide, CO | 26.0 |
| Hydrogen, H | 18.5 |
| Marsh gas, $CH_4$ | 0.5 |
| Nitrogen, N | 47.0 |
| Heat value, cu. ft. | 145 |

This gas is sometimes called semi-water gas, owing to the fact that steam is usually jetted into the suction pipe.

## Liquid Fuels.

The liquid fuels, derived from crude petroleum by successive distillation, are gasoline, naphtha, kerosene, gas-oil, and fuel oils. Alcohol, derived from the fermentation and distillation of organic matter, is being largely experimented with as a probable substitute for the petroleum products, which, owing to the decrease of the visible supply and to trade conditions, seems destined to increase in price, while the probable tendency of alcohol will be to decrease in price. While the rate of flame propagation in alcohol vapor is slower than in that of gasoline or benzene, it is claimed, and the claim is substantiated by government tests, that the mean effective pressure obtained by the use of alcohol equals or exceeds that of gasoline under the same conditions, providing that the apparatus for vaporizing the fluid is suited to the requirements. As we shall subsequently show that the temperature required to secure a mixture most suitable for complete combustion is considerably higher for alcohol than for the lighter petroleum products, it is necessary that more heat be supplied in the vaporizer or carburettor in order to obtain these results.

Table III gives the approximate percentage of the successive distillates of petroleum, with their specific gravities, boiling points, and flashing points. The values are subject to considerable variation and should not be taken as more than averages.

### TABLE III.

PROPERTIES OF PETROLEUM DISTILLATES.

| Name of Product. | Per cent. | Sp. Gr. | Boiling Point. Deg. Fahr. | Flashing Point. Deg. Fahr. |
|---|---|---|---|---|
| Rhigolene | Trace | 0.595 | 113 | ........ |
| Chimogene | Trace | 0.625 | 130 | ........ |
| Gasoline } Commercial gasoline. | 0.02 | 0.650 | 125 | 10 |
| Benzene, naphtha } Commercial gasoline. | 0.10 | 0.700 | to | 14 |
| Kerosene, light } Commercial gasoline. | 0.10 | 0.730 | 200 | 50 |
| Kerosene, medium | 0.35 | 0.800 | 400 | 150 |
| Kerosene, heavy | 0.10 | 0.890 | 550 | 270 |
| Lubricating oil, heavy | 0.10 | 0.873 | ........ | 315 |
| Cylinder oil | 0.05 | 0.915 | ........ | 360 |
| Vaseline | 0.02 | 0.925 | ........ | ........ |
| Waste products | 0.16 | ........ | ........ | ........ |
| | 1.00 | ........ | ........ | ........ |

### TABLE IV.

COMPOSITION OF CRUDE OILS.

| Constituent. | Beaumont. | Penn'a.* | Ohio.† |
|---|---|---|---|
| Carbon | 85.03 | 86.10 | 85.00 |
| Hydrogen | 12.30 | 13.90 | 13.80 |
| Sulphur | 1.75 | 0.06 | 0.60 |
| Oxygen and hydrogen | 0.92 | ........ | 0.60 |
| Loss on treatment with excess of $H_2SO_4$ (sulphuric acid) | 39.00 | 21.00 | 30.00 |

* Engler. † Mabery, Noble County.

Table from "Oil Engines," Goldingham.

The heating value of this oil, per pound, is about 20,500 B.t.u.

## Gas Oil.

Gas or fuel oil is procured, in the process of fractional distillation, after the products benzene, gasoline, and kerosene have been driven off. The oil is much used in oil engines, but with the heavier oils a carbon deposit may be expected in the cylinder, unless suitable means are provided for vaporizing and heating the fuel. There is little, if any, difference between fuel oil and gas oil, both being made up of various constituent parts, are subject to wide variation in their value as a fuel, and unless an oil is known to be suitable for use in a gas engine it should be analyzed. The heating value in heat units per pound is usually taken as slightly lower than crude oil, that is, about 18,000 per lb.

## Gasoline.

Commercial gasoline varies in specific gravity from 0.65 to 0.74, the variation being due to the varying amounts of hydrogen in its composition. Gasoline vapor is equal to about 160 cu. ft. per gal. or a little more than 1,100 times its liquid bulk. The following is the ultimate composition, by test, of 0.71 specific gravity gasoline as used in government tests, recently run, on alcohol and gasoline fuel.

| | |
|---|---|
| Carbon | 85 |
| Hydrogen | 14.8 |
| Not named | 0.2 |
| | 100.0 |

Table V, also taken from the results of government experiments, "Tests of Internal-Combustion Engines on Alcohol Fuel," Bulletin 191, is the result of fractional distillation of a sample of gasoline. The range of temperatures show conclusively the complex mixture of this fuel.

TABLE V.

FRACTIONAL DISTILLATION OF GASOLINE.

| Number of Fraction. | Temperatures. Deg. Fahr. | Number of Fraction. | Temperatures. Deg. Fahr. |
|---|---|---|---|
| 1 | 114.8 to 140.0 | 8 | 206.6 to 212.0 |
| 2 | 147.2 to 167.0 | 9 | 212.0 to 219.2 |
| 3 | 167.0 to 176.0 | 10 | 219.2 to 226.4 |
| 4 | 176 constant. | 11 | 226.4 to 233.6 |
| 5 | 176.0 to 186.8 | 12 | 233.6 to 248.0 |
| 6 | 186.8 to 197.6 | 13 | 248.0 to 258.8 |
| 7 | 197.6 to 206.6 | 14 | 258.8 to 284.0 |
| | | 15 | 284.0 to 311.0 |

At the final temperature there were still 5 cu. cm. not vaporized.

A lighting gas composed of about equal parts of air and gasoline vapor, about 320 cu. ft. being obtained from a gallon of fuel, is not explosive and is used quite extensively as an illuminant. The gasoline is stored in an outdoor tank or carburettor, and the pressure, equivalent to that due to from 1 in. to 1.5 in. of water, is obtained by an air pump actuated by means of a weight or suitable motor. The operating principle of the pump is an inversion of the wet gas meter principle. A mixing valve automatically controls the mixture by means of a balance obtained between the atmosphere and the mixture of gasoline vapor and air. This illuminating gas, when properly mixed with air, becomes a good fuel for gas engines.

Seventy-one test gasoline weighs 5.91 lb. per gal.; about 27 cu. ft. of its vapor is obtainable per pound and $\frac{19{,}000}{27} = 703$ heat units per cu. ft.

## Kerosene.

Kerosene oil has a specific gravity of from 0.77 to 0.82 and weighs, for 0.80 test, 6.7 lb. per gal. It has a heat value a trifle higher than that of gasoline, about 22,000 B.t.u. per lb. The oil ignites when heated to from 130° to 140° fahr. and flashes at about 120° fahr.

## Heat of Combustion.

The heat of combustion of different fuels is obtained by means of a calorimeter, in which the heat generated by a known amount of fuel is absorbed by a given weight of water. The standard of measurement in this country is the B.t.u., which is the heat required to raise the temperature of one pound of water from 39° to 40° fahr. In European countries, where the centigrade thermometer is used, heat of combustion is measured in calories per kilogram. Heat of combustion in calories is five-ninths that of the heat of combustion in B.t.u.

The heat of combustion of the petroleum oils varies from 18,000 to 22,000 B.t.u. per lb. and 20,000 is an average value. The value for pure ethyl alcohol, as obtained from experiment, is about 12,700, or approximately two-thirds that of the petroleum oils; as experiments have shown that ethyl alcohol as a fuel has about two-thirds the value of gasoline for the same weight, these figures are significant as serving to show that the relative thermal efficiencies are about equal.

All liquid fuels contain hydrogen in greater or less amount, which when burned is converted into water vapor. When the fuel is tested in a calorimeter, this steam, coming in contact with the walls of the calorimeter, is condensed and contributes quite an appreciable amount of heat to the water of the calorimeter. Since the products of combustion in a gas-engine cylinder always leave the cylinder at a temperature considerably above that of boiling water, the engine does make use of the latent heat of condensation, which the calorimeter measures, and, in consequence, the effective heat value of the fuel, when burned in the cylinder of an engine, is of a value lower than the theoretical as obtained by calorimeter test. For this reason it is customary to compute, and deduct, this latent heat of condensation when comparing engine fuels. The resulting value obtained is known as the low heat value. Supposing the heat value of gasoline, as obtained by calorimeter test, was 21,000 B.t.u.

Taking the chemical symbol to be $C_6H_{14}$, the percentage of H, by weight, would be $\frac{6 \times 12}{14 \times 1} = 0.194$, and since hydrogen unites with oxygen to produce water, in the ratio of 1 to 8, by weight, then the weight of water (vapor) produced in the combustion of 1 lb. of gasoline would be $(0.194 \times 8) + 0.194 = 1.746$, and the latent heat of water at atmosphere being found, from the steam tables, to be 965.5°, then, for 1.746 lb. the latent heat would be $1.746 \times 965.5° = 1680°$, and $21{,}000 - 1680 = 19{,}320$ B.t.u., the low heat value for gasoline of the above chemical composition. Since the chemical composition of gasoline is subject to such wide ranges, the actual low heat value can only be obtained by securing its volumetric analysis, to determine the actual amount of hydrogen in its composition, and then proceeding as above.

### Air Necessary for Combustion.

The air required for the complete combustion of a fuel of definite known composition can be accurately determined. This calculation may be made for a fuel such as alcohol, but can only be approximately determined for the petroleum fuels.

For ethyl alcohol we have the formula $C_2H_5OH$. Its molecular weight is:

| | |
|---|---|
| Carbon | $2 \times 12 = 24$ |
| Hydrogen | $6 \times 1 = 6$ |
| Oxygen | $1 \times 16 = 16$ |
| | 46 |

For the complete combustion of one molecule of alcohol, the two atoms of carbon require four atoms of oxygen to produce $CO_2$ (carbon dioxide) and the six atoms of hydrogen require three atoms of oxygen to produce $H_2O$ (water vapor). There is one atom of oxygen in the molecule of alcohol and hence the oxygen which must be derived from outside sources is $(3 + 4) - 1 = 6$ atoms. The weight of the 6 atoms is $6 \times 16 = 96$. Hence complete combustion of 1 lb. of $C_2H_5OH$ requires $\frac{96}{46} = 2.086$ lb. of oxygen, and since in 1 lb. of dry air there is 0.230 lb. of oxygen, to obtain 2.086 lb. would require $\frac{2.086}{0.230} = 9.06$ lb. of air. If the air is quite damp, one pound would contain less free oxygen and hence more air would be required.

Assuming the composition of gasoline to be $C_6H_{14}$, the amount of air required for complete combustion would be determined in like manner. The molecular weight is:

| | |
|---|---|
| Carbon .................................... | $6 \times 12 = 72$ |
| Hydrogen .................................. | $14 \times 1 = 14$ |
| | 86 |

The 6 atoms of carbon require 12 atoms of oxygen and the 14 atoms of hydrogen require 7 atoms of oxygen. Now since all must be derived from outside sources the amount of oxygen required is $12 + 7 = 19$. The molecular weight of the oxygen would be $19 \times 16 = 304$, and $\frac{304}{86} = 3.534$ lb. of oxygen per pound of gasoline, and $\frac{3.534}{0.230} = 15.37$ lb. of air required.

Table VI gives the amount of air required for the combustion of one pound of different fuels, assuming their chemical formulas to be correct.

TABLE VI.

AIR REQUIRED FOR COMBUSTION OF DIFFERENT FUELS.

| Substance. | Formula. | Pounds Air. | Cu. Ft. Air. 60° Fahr. |
|---|---|---|---|
| Alcohol, ethyl .................... | $C_2H_5OH$ | 9.06 | 118.57 |
| Alcohol, methyl .................. | $C_2H_4O_2$ | 4.63 | 60.60 |
| Gasoline............................ | $C_6H_{14}$ | 15.37 | 201.17 |
| Gasoline vapor.................... | $C_2H_{14}$ | 25.91 | 339.13 |
| Kerosene............................ | $C_{10}H_{22}$ | 6.52 | 85.34 |
| Benzene............................. | $C_6H_6$ | 13.34 | 174.60 |
| Acetylene........................... | $C_2H_2$ | 13.34 | 174.60 |
| Marsh gas (methane)............ | $CH_4$ | 17.39 | 227.61 |
| Olefiant gas (ethylene)........... | $C_2H_4$ | 14.87 | 194.63 |

The mixture of water with the alcohol reduces the amount of air necessary for complete combustion. The ordinary commercial denatured alcohol generally carries 10 per cent of water.

## Vaporization.

In order that a liquid fuel may be used in an internal-combustion engine, it must first be vaporized, and the vapor thus obtained mixed with air in the proper proportion.

For their complete vaporization, fuels, such as kerosene, with a high boiling point, require much greater heating in the vaporizer, than fuels with a comparatively low boiling point. Alcohol, kerosene, and the crude oils require greater heat than the lighter hydrocarbons. Gasoline, being extremely volatile, is easily vaporized at ordinary temperatures.

All substances which liquefy at ordinary temperatures have a definite limit to the amount of their vapor which may exist in any given space at a given temperature. According to the laws for perfect gases, which we assume to hold true in the case of these vapors, at a given constant temperature the weight of any vapor present in a cubic foot of space is proportional to its vapor pressure and may be measured by it. Illustrating this, imagine a cylinder containing any vapor, at a pressure corresponding to 20 mm. of mercury, and maintained at a constant temperature of 60° fahr., during the experiment, to be fitted with a perfectly tight piston. If the vapor is now compressed, by means of the piston, to half its original volume, the pressure will rise to 40 mm. of mercury, and if this volume is again halved, its pressure will rise to 80. But for any vapor there is a saturation pressure corresponding to every degree of temperature, and if the compression is carried beyond this point the pressure will not rise but will remain constant and, as the volume is further decreased, a part of the vapor will be condensed into a liquid, but the amount of vapor per cubic foot of space will remain constant. Suppose, for illustration, that the vapor being compressed was that of gasoline, and that we carried the pressure to that of 160 mm. of mercury. We would then find that a part of the vapor had begun to condense, as the vapor pressure of saturation for gasoline vapor at 60° fahr. is about 157.

Table VII gives the vapor pressure of saturation for several substances as found in the Bulletin 191, United States Department of Agriculture.

TABLE VII.

VAPOR PRESSURE OF SATURATION FOR VARIOUS SUBSTANCES.

| Temperature. | | Vapor Pressure of Saturation in mm. of Mercury. | | | |
|---|---|---|---|---|---|
| Deg. C. | Deg. F. | Pure Ethyl Alcohol. | Pure Methyl Alcohol. | Water. | Gasoline. |
| 0 | 32 | 12 | 30 | 5 | 99 |
| 5 | 41 | 17 | 40 | 7 | 115 |
| 10 | 50 | 24 | 54 | 9 | 133 |
| 15 | 59 | 32 | 71 | 13 | 154 |
| 20 | 68 | 44 | 94 | 17 | 179 |
| 25 | 77 | 59 | 123 | 24 | 210 |
| 30 | 86 | 78 | 159 | 32 | 251 |
| 35 | 95 | 103 | 204 | 42 | 301 |
| 40 | 104 | 134 | 259 | 55 | 360 |
| 45 | 113 | 172 | 327 | 71 | 422 |
| 50 | 122 | 220 | 409 | 92 | 493 |
| 55 | 131 | 279 | 508 | 117 | 561 |
| 60 | 140 | 350 | 624 | 149 | 648 |
| 65 | 149 | 437 | 761 | 187 | 739 |

Gasoline, being an extremely volatile substance, has a comparatively high saturation pressure, while for the alcohols it is comparatively low.

Since the vapor pressure determines the amount of vapor which may be contained in any given volume for a given temperature and pressure, if we know the amount of vaporous fuel necessary to produce the best explosive mixture, when mixed with air, we may determine the lowest temperature at which this may exist at atmospheric pressure. As the vapor pressure of alcohol is low, we may expect to find the necessary temperature correspondingly high. The existence of different gases or vapors in the same receptacle, if they have no chemical affinity for each other, does not affect their action. Then a mixture of dry air and alcohol vapor, or vapor of a hydrocarbon product, in the proper proportions, as previously determined, for complete combustion, will, when under pressure, derive a portion of its vapor pressure from the air and a portion from the vapor of the fuel in the mixture. By means of Avogadro's law we are able to compute the vapor pressure due to the vaporized fuel. *Pressure and temperature remaining constant, the density of any gas is proportional to its molecular weight.* Hydrogen gas is taken as the basis of comparison. Since the molecular weight of hydrogen is 2 and that of ethyl alcohol is 46, then the relative density is $\frac{46}{2} = 23$. That is to say, 1 lb. of ethyl alcohol vapor confined in a given space has a vapor pressure equal to one-twenty-third that of hydrogen confined in the same space and at the same constant temperature. The density of air being 14.44, on the same basis of comparison, then 9.06 lb. of air, the amount, as previously determined, necessary for complete combustion, would have $\frac{1}{14.44}$ part of 9.06 of the vapor pressure of 1 lb. of hydrogen. Then the relative vapor pressures of the alcohol vapor and air are as $\frac{1}{23} : \frac{9.06}{14.44}$ or as 0.0435 : 0.627. The total vapor pressure may then be considered as equal to $0.0435 + 0.627 = 0.670$, of which $\frac{0.0435}{0.670} = 6.5$ per cent is due to the alcohol vapor, and 93.5 per cent is due to the dry air. Under ordinary atmospheric pressure of 760 mm. of mercury, that is, 14.7 lb. per sq. in., the vapor pressure of the alcohol would be $760 \times 0.065 = 49.$ mm., which we find from the table corresponds to a temperature of a trifle more than 71° fahr. Hence, a perfectly combustible mixture of air and alcohol cannot exist at a temperature of less than 71° fahr., although a mixture carrying an excess of air may exist at a much lower temperature. If the air is damp and the alcohol, as is nearly always the case, carries a certain per cent of water, then the temperature must be higher, as the air is already partially saturated, due to the pressure of water vapor.

All this shows conclusively that alcohol to be changed from a liquid to a vapor must receive a relatively large amount of heat, either in the mixing valve or carburettor, by direct contact with heated surfaces, or in the air supplied for the mixture. Too much heat will produce too rich a mixture, and result in incomplete combustion, as well as lowering the heat efficiency of the engine.

Now considering a mixture of gasoline vapor and air of the same relative humidity as used for the alcohol mixture. We will consider that its chemical formula is $C_6H_{14}$, and as we have previously determined the amount of air necessary for complete combustion is 15.37 lb., then $\frac{1}{43} : \frac{15.37}{14.44}$ as the vapor pressure of the gasoline vapor is to that of air. (The molecular weight of the gasoline is 86.) In terms of decimals the relation is as 0.0233 : 1.064. The total vapor pressure being considered as equal to $0.0233 + 1.064 = 1.0873$, the per cent due to the gasoline vapor is $\frac{0.0233}{1.0873} = 2.14$ per cent, and the per cent due to the air is 97.86 per cent. Now $760 \times 0.021 = 15.96$ mm., which we find from the table corresponds to a temperature of considerably less than 32° fahr. This latter computation can only be considered as an approximation because of the different compositions of gasoline, but it serves to show the comparative ease with which gasoline may be vaporized and at any ordinary temperature. Similar computations may be made for any one of the fuels used in internal-combustion engines, if the chemical composition is known.

### Acetylene.

Some experiments have been made in the use of gas derived from the action of water on calcium carbide. This gas is known as acetylene and is quite extensively used as a lighting gas.

Acetylene liquefies, at atmospheric pressure, at a comparatively high temperature, about $-116°$ fahr., while under a compression of 600 to 700 lb. it liquefies at temperatures higher than the ordinary temperatures of air. It is suggested by some that liquid acetylene would be a very available fuel for internal-combustion engines.

Acetylene has a heating value of about 18,000 B.t.u. per lb., its formula being $C_2H_2$, and it requires 10.78 lb. of air for the complete combustion of one pound of gas.

One pound of calcium carbide will yield about 0.35 lb. of the gas and will use about 0.5 lb. of water in the process.

One pound of liquid acetylene, evaporated, will produce 14.5 lb. of gas at atmospheric pressure.

Acetylene gas ignites spontaneously at lower pressures than gasoline vapor or natural gas, and, in consequence, is not available for high-compression engines.

Results of foreign experiments seem to show that it requires from 5 to 7 cu. ft. of acetylene, mixed with air in the proper proportion, to produce 1 hp.-hr.

The rate of flame propagation is high, the temperature of combustion is high, and the energy derived is high, but the cost of production of the gas is more than twice that of gasoline vapor, for the same amount of work. The liberation of the gas, in a closed vessel, may also produce a dangerously high pressure. It is doubtful if the use of acetylene as a fuel will become very general unless the cost of production is considerably lessened or the cost of the other available fuels increased to such an extent as to make its use comparatively economical.

## Alcohol.

We have already shown the action of alcohol, in regard to its vaporization and combustion, to be considerably different from that of gasoline. The low heat value of denatured alcohol is about 11,000 B.t.u., while that of commercial gasoline is slightly more than 18,000; it is thus apparent that their heating values are approximately as 2 : 3, or, in other words, the thermal value of alcohol is a trifle less than two-thirds that of gasoline. This fact is significant when we consider that the average of the government tests on the comparative values of gasoline and alcohol shows that the ratio of their consumption per b. hp. in the same engine and under the same conditions was approximately as 0.64 : 1.00. It seems to show that the operation of alcohol and of gasoline in an internal-combustion engine is practically the same as regards the thermal efficiency. It is possible, however, with alcohol, to use a higher compression than with gasoline, and earlier ignition may be used and still produce a smooth-running engine.

A gasoline engine of ordinary design may burn alcohol with more or less success, but to secure the best results the carburettor must be adapted to the requirements of alcohol vaporization. When properly arranged an engine will deliver slightly more power with alcohol than with gasoline. Alcohol being of known chemical composition, unmixed with impurities other than water, has no inherent tendency to foul the interior of the cylinder.

The exhaust from a gasoline engine is much hotter than that of an engine operating on alcohol, and, in consequence, there is less danger from fire in the latter.

Alcohol may be used and stored with much less danger than gasoline, and its use, both in the matter of the exhaust products and in the handling, is much more pleasant. Alcohol can be produced anywhere from the distillation of organic waste products.

The cost of alcohol is still much more than that of gasoline, and unless the price of the latter increases considerably or the price of alcohol is decreased, it is doubtful if it will come very rapidly into general use.

# CHAPTER X.

## COMPRESSION.

THE compression pressure of an internal-combustion engine varies from an extreme high limit of 500 lb. per sq. in. or a trifle more to 60 lb. per sq. in., both these pressures being above atmosphere. As the pressure is increased, the temperature of the compressed gases increases until with a pressure of 500 lb. the gas attains a temperature of nearly 1000° fahr. and is incandescent.

The compression in any type of engine is governed entirely by the conditions. If a fuel with a high ignition point be used, the compression may be high, as in the use of producer gas, in which engines the compression is carried as high as from 150 to 200 lb. per sq. in. In the Diesel type of motor it is possible to use the extremely high compression of 500 lb. per sq. in., owing to the fact that during the compression stroke the gas in the cylinder is pure air, and, as previously described, ignition is obtained by the injection of the fuel into this incandescent cylinder. The compression of a low-speed gasoline engine should not exceed 85 lb. per sq. in., as this produces a temperature in the cylinder sufficient in some cases to ignite the charge. As the engine speed increases, the compression may be increased to such a point that ignition, due to the compression, does not take place previous to the proper lead of ignition, in the compression stroke, for the speed in question. Many automobile engines are built with a compression as high as 95 lb. gage, and give excellent satisfaction when operating at an average speed of from 800 to 900 rev. per min., while a two-cylinder engine which operated at a speed of from 300 to 400 rev. per min. and with the same compression gave constant trouble from premature ignition.

The compression in common use in gas engines operating with natural gas as a fuel varies from 60 to 120 lb. gage pressure, and as a rule the compression space of a gas engine does not vary much from that of a gasoline engine of the same speed. However, if gas and gasoline vapor ignited at the same temperature, which is not the case, the ignition point of the latter being considerably lower, the fact that gasoline vapor burns more rapidly would permit of less lead of the ignition in the compression stroke and, in consequence, a lower compression would be necessary. Taking into consideration, then, (1) that gasoline vapor ignites at a lower temperature and (2) that it burns more rapidly, we may safely design a natural gas engine with a compression from 10 to 15 lb. higher than for a gasoline engine of like speed.

TABLE VIII.

COMPRESSION TEMPERATURES.

| Absolute Pressure in Lb. per Sq. In. | Temperature of Compression. Fahr. | Absolute Pressure in Lb. per Sq. In. | Temperature of Compression. Fahr. |
|---|---|---|---|
| 14.7 | 59.9 | 80.0 | 390.67 |
| 15.0 | 63.06 | 85.0 | 405.79 |
| 20.0 | 108.68 | 90.0 | 418.48 |
| 25.0 | 148.56 | 95.0 | 434.21 |
| 30.0 | 179.78 | 100.0 | 447.64 |
| 35.0 | 209.08 | 125.0 | 508.39 |
| 40.0 | 236.95 | 150.0 | 561.04 |
| 45.0 | 259.77 | 175.0 | 607.77 |
| 50.0 | 282.13 | 200.0 | 649.96 |
| 55.0 | 302.95 | 225.0 | 688.48 |
| 60.0 | 322.48 | 250.0 | 772.83 |
| 65.0 | 340.90 | 300.0 | 788.61 |
| 70.0 | 358.32 | 400.0 | 897.37 |
| 75.0 | 374.88 | 500.0 | 988.18 |

Compression in oil engines operating with hot-tube or electric ignition devices may be carried to 85 or 90 lb. gage per sq. in., and possibly somewhat higher, the compression depending to a large extent on the grade of fuel used. Several oil engines, notably the Mietz & Weiss and the Hornsby-Akroid, obtain their ignition by means of high compression and residual heat in the vaporizer. These engines compress their charge to about 200 to 210 lb. per sq. in.

The efficiency of an internal-combustion engine increases with its compression, as may be shown.

We have the two equations for the heat absorbed at constant volume and constant pressure.

$$\left.\begin{aligned} dH &= C_v dt + pdv \\ dH &= C_p dt - vdp \end{aligned}\right\} \quad . \qquad (1)$$

Supposing the gas to expand without the transmission of heat

$$dH = 0.$$

$$\therefore\ C_v dt = -pdv \qquad (a)$$

$$C_p dt = vdp \quad . \quad . \quad . \quad . \qquad (b)$$

Dividing $(b)$ by $(a)$ gives

$\frac{dp}{p} = -\frac{C_p}{C_v} \cdot \frac{dv}{v} = -\frac{\gamma dv}{v}$ ($\gamma$ being the ratio of specific heat at constant pressure to the specific heat at constant volume).

Integrating we have

$$\log_e \frac{p_y}{p_x} = \log_e \left(\frac{v_x}{v_y}\right)^\gamma .$$

In which $p_x$ and $v_x$ are the initial limits and $p_y$ and $v_y$ are the general limits.* Then

$$p_y v_y^\gamma = p_x v_x^\gamma = C \quad . \quad . \qquad (2)$$

Now a perfect gas is one in which the rate of expansion equals the rate of temperature increase if under constant pressure, or the volume being constant, the temperature will increase as the pressure. Then

$$p_y v_y \propto t_y \text{ and } p_x v_x \propto t_x .$$

* In an indicator diagram the limits $p_x$ and $v_x$ correspond to the pressures and volumes at the beginning of the compression or the end of the exhaust stroke.

We may now eliminate $p_x$ and write equation (2),

$$\frac{t_y}{t_x} = \left(\frac{v_x}{v_y}\right)^{\gamma-1} = \left(\frac{p_y}{p_x}\right)^{\gamma-1} \quad . \quad . \quad . \quad . \quad . \quad (3)$$

Now, supposing $v$ to be a constant, we would have from equation 1, since the derivative of a constant is 0,

$$H = C_v\,(t_x - t_y) \quad . \quad . \quad . \quad . \quad . \quad (4)$$

(The heat absorbed at constant volume is equal to the specific heat at constant volume times the range of temperature.)

Drawing the theoretical card 1, 2, 3, 4 (Fig. 25), in which heat is considered as being absorbed from 4 to 1 and rejected

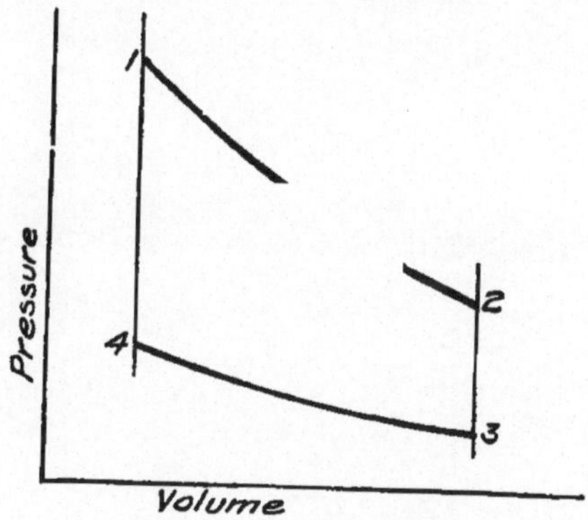


**Fig. 25. Theoretical Card.**

from 2 to 3, the constant volume lines, 4–1 and 2–3, being the theoretically perfect explosion and release lines in an internal-combustion engine, the efficiency would be

$$E = \frac{H_{1-4} - H_{2-3}}{H_{1-4}} = \frac{t_1 - t_4 - (t_2 - t_3)}{t_1 - t_4} = 1 - \frac{(t_2 - t_3)}{(t_1 - t_4)}.$$

But $v_1 = v_4$ and $v_2 = v_3$, then if $\gamma$ be constant, we have, from equation 3,

$$\frac{t_2}{t_1} = \frac{t_3}{t_4}.$$

$$\therefore \frac{t_2 - t_3}{t_1 - t_4} = \frac{t_2}{t_1} = \frac{t_3}{t_4},$$

$$\text{and } \therefore E = 1 - \frac{t_2}{t_1} = 1 - \frac{t_3}{t_4} = 1 - \left(\frac{p_3}{p_4}\right)^{\gamma-1}.$$

Now as the compression is increased, the value of the fraction, $\frac{p_3}{p_4}$, will decrease, since the initial pressure $p_3$ is that of the atmosphere, and $p_4$ increases. The value of $E$ will then approach 1, as a limit. If it were possible for $E$ to equal 1, then the efficiency would be the maximum theoretical value, which value is never even approached in a heat engine.

Equation 2 is that of the ideal indicator card and its application will be fully discussed in the following chapter.

# CHAPTER XI.

## THE INDICATOR CARD.

As the indicator card of an engine, as obtained by means of an actual indicator test, is a record, more or less perfect, of the operations taking place within the cylinder, the question arises as to what a perfect cycle in that particular cylinder would constitute, and, knowing the operations of this theoretically perfect cycle, it is the attempt of the designer to approximate, as nearly as may be, this ideal condition.

The theoretical indicator card is obtained for any particular engine by the use of one general formula, which has as its basis the known action of gases under compression or expansion. By means of this formula it is possible to compute a number of values, representing the pressures in the cylinder at different points in the cycle, and having obtained these values, to plot them or to draw the curves of compression or expansion.

In the design of the cams the points at which the exhaust and inlet valves open and close have been determined and, in consequence, their operation will affect the card at some known point.

In Chapters I, II and III, describing the several cycles, the cards are shown in their exact relation to the stroke of the piston. In Fig. 26, the stroke may be represented by *bc*, and the piston clearance, or compression space, by *ab*. As in Chapters I, II and III, the stroke is shown divided into ten equal parts by ordinates 1 to 10. By means of the indicator card formula, derived in the chapter on "Compression," we will be able to compute the values for the points on the curves, *cd* and *ef*, intersected by these ordinates. The general form of equation for the compression curve *cd* and also for the expansion curve *ef* is

$$pv^{\gamma} = k \quad (5)$$

In which

$p$ = pressure above a vacuum.
$v$ = volume at pressure $p$.
$k$ = constant depending upon conditions *but which will remain the same for all points of the same curve.* For convenience we will denote the constant for the compression curve by $k_c$ and that for the expansion curve by $k_e$. Considering the volume of the cylinder as unity, the value of the constant $k_c$ will be the pressure of the atmosphere, which we assume to be the pressure in the cylinder at the beginning of the compression stroke and which is taken to be 14.7 lb. per sq. in. The determination for the value of $k_e$ takes into consideration the fact that the pressure of the explosion is approximately four times the compression, and, having this value determined, we may solve the formula for $k_e$, which will remain constant throughout the expansion stroke and is the value of $p$ if the gases were to release at the end of the stroke, $g$.

If a volume of air is contained in the perfectly tight cylinder fitted with a perfectly tight piston and compressed at constant temperature, then the value of $\gamma$ would become 1. If the volume of air is rapidly compressed and all the heat produced by this compression is retained, then the value of $\gamma$ will increase. The value of $\gamma$ is the ratio between the specific heat of air, or whatever gas is under compression, at constant pressure and its specific heat at constant volume. According to Rankine,

$$\gamma = \frac{S_p}{S_v} = 1.408 \text{ for air.}$$

For the gases in common use in the cylinder of the gas engine the value of $\gamma$ for the compression and expansion curve is considerably lower, due to the ratio of the specific heats being different from those for air and to the losses due to leakage past the piston and valves and the loss of heat through the cylinder walls. For general practice the values of $\gamma$ may be taken as 1.3 for the compression curve and 1.35 for the expansion curve, and these values will be found to produce results very closely in keeping with the actual working pressures obtained by indicator tests on efficient types of internal-combustion engines.

The value 1.3 is very nearly the average of results obtained by computation of $\gamma$ for different samples of natural gas; so

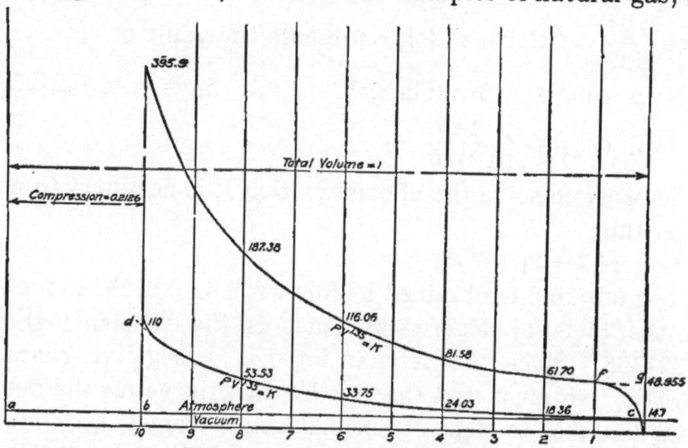


Fig. 26. Ideal Indicator Card.

it may be said that the average internal-combustion engine is designed on the basis of natural gas as a fuel and acting under ideal conditions.

The value $\gamma = 1.35$ is taken for the expansion curve in view of the fact that the higher temperature of the gases during the working stroke results in a greater difference in temperature between the contents of the cylinder and the jacket water, with a consequent greater loss of heat.

In order to show just how the ideal indicator card may be built up by means of the formulas, we will compute the values for several points of the compression and expansion curves in Fig. 26.

Before laying out the diagram or computing any values, it is necessary to decide upon some pressure for the compression or to assume some ratio for the volume of the compression space to that displaced by the piston during one stroke, and to solve for the compression pressure. In this model card we will assume the ratio to be as 0.27:1, which will result in a rather high compression pressure. The total volume of the cylinder at the end of the suction stroke then will be $1 + 0.27 = 1.27$ times the piston displacement. For convenience in the computations we will assume this volume as unity. The percentage of the compression space to the total volume of the cylinder is then $\frac{0.27}{1.27} = 0.2126$, which represents the value of $v$.

Now using the formula $pv^{1.3} = k$, we have $p \times 0.2126^{1.3} = 14.7$ or $p = \frac{14.7}{(0.2126)^{1.3}}$.

In order to solve the above equation it is necessary to use logarithms.

Log 14.7 = 1.167317.

Log $0.2126^{1.3}$ is obtained as follows: Log 0.2126, as found in the table of logarithms (that is the tabular log), is $\bar{1}.327563$. Since the tabular log has a negative characteristic,* we must find the true log by subtracting the mantissa (number to right of decimal point) from 1, as follows.

$\bar{1}.327563$ may be written 9.327563–10.

Then we may subtract

```
10.000000-10
 9.327563-10
 0.672437       True log 0.2126
      1.3       Multiplying by 1.3 to raise to that power
  2017311
  672437
 0.874168       True log 0.2126^1.3
 1.125832       Tabular log 0.2126^1.3
Now log 14.7 - log 0.2126^1.3 = log p.
11.167317-10    log 14.7
 9.125832-10    log 0.2126^1.3
 2.041485       Tabular log p.
```

* The characteristic is the number to the left of the decimal point and is 1 for any number in tens place, as 16.53; 2 for any number in hundreds place, as 165.32; 3 for any number in thousands place, as 1653.2, etc. For decimals, the characteristic is: – 1 for any number in tenths, as 0.165; – 2 for any number in hundredths, as 0.0165; – 3 for any number in thousandths, as 0.00165, etc.

From the table we find the number corresponding to the logarithm 2.041485 to be 110.025, which is the value of $p$ in lb. per sq. in., above a vacuum. For all practical purposes it would be sufficiently close to assume $p = 110$. (This value of $p$ is in common use in the high-compression automobile engines, but is too high for general use in most stationary engines operating on natural gas. Engines operating on poorer quality of gas, as producer gas, may use a much higher compression.)

In order to obtain the true outline of the compression curve it is necessary to solve for several other points on the curve. The points for which the calculations will be made, in this instance, are those where the piston is at 0.2, 0.4, 0.6, and 0.8 of the compression stroke and the volumes are respectively $\frac{1.27 - 0.2}{1.27} = 0.8425$, $\frac{1.27 - 0.4}{1.27} = 0.685$, $\frac{1.27 - 0.6}{1.27} = 0.5276$, and $\frac{1.27 - 0.8}{1.27} = 0.37$, which points will generally prove sufficient for any compression curve. The corresponding pressures are obtained by the same formula, making proper allowance for the increased volumes, and are found to be 18.36, 24.03, 33.75, and 53.53, respectively.

The pressures for the expansion curve may now be determined. The compression being 110 lb. per sq. in. absolute, then the approximate explosion pressure may be found as follows: $110 - 14.7 = 95.3$ (pressure above atmosphere). Then $4 \times 95.3 + 14.7 = 395.9$ (explosion pressure). By application of the formula we are able to obtain the value of the constant $k_e$:

$$pv^{1.35} = k_e$$

$$\text{or } 395.9 \times 0.2126^{1.35} = k_e$$

$$\log 395.9 = 2.597585$$

$$\log 0.2126^{1.35} = \bar{1}.092210$$

$$\log k_e = 1.689795$$

Then $k_e = 48.955$ lb., the terminal pressure in the cylinder considering the exhaust valve to open at the end of the working stroke.

Hence the equation for any point of the expansion curve in question is $pv^{1.35} = 48.955$. The intermediate working pressures, corresponding to the computed compression pressures, at 0.2, 0.4, 0.6, and 0.8 of the working stroke, are found by subtracting the logs of the different values of $v$, as previously determined, raised to the 1.35th power from the log of 48.955. They are 61.70, 81.58, 116.06, and 187.38.

Sufficient points now having been determined, the curves are drawn as shown in Fig. 26. The scale of indicator spring in this diagram is taken to be 140 lb. per in. and the length of the diagram from *a* to *c* is 4 in. Then every inch in height on the diagram represents 140 lb. pressure, and unit volume is represented by the length 4 in.

We now have completed the theoretical diagram *cdeg*, which would represent the action in the cylinder if the exhaust valve opened at *g* and the pressure dropped on a vertical line to *c*. Plainly this would be purely an ideal condition. To complete the actual card it becomes necessary to pass a curve from the intersection, *f*, of ordinate 1 and the expansion curve, at which point the exhaust is supposed to open, to point *c*. This curve should be the arc of a circle, having its center on ordinate 1 and intersecting points *f* and *c*. At *d* a slight curve, representing the increase of pressure due to the advance of the ignition, is drawn. A straight vertical line drawn from the intersection of this curve and ordinate 10 to point *e* completes the diagram.

The diagram having been drawn, the mean effective pressure may be computed as described in Chapter XXIII. The mean effective pressure having been found, the designer may compute the horsepower of any proposed engine of like compression, or he may determine the dimension of bore and stroke to use for any proposed horsepower.

Fig. 27, representing the compression curve $pv^{1.3} = k$, may be used to determine the compression space and to locate any point on the compression curve from 20 up to 500 lb. By the use of this curve the computation of the values for points on the compression curve is rendered unnecessary. Fig. 27 also shows the terminal pressures of the expansion curve for compressions from 60 to 120 lb., all these values being computed on the assumption that the absolute explosion pressure is four times the absolute compression pressure.

To use the curve proceed as follows: If the compression has been determined and the per cent of compression space is required, find, on ordinate $P$ or $P_1$, the pressure of com-

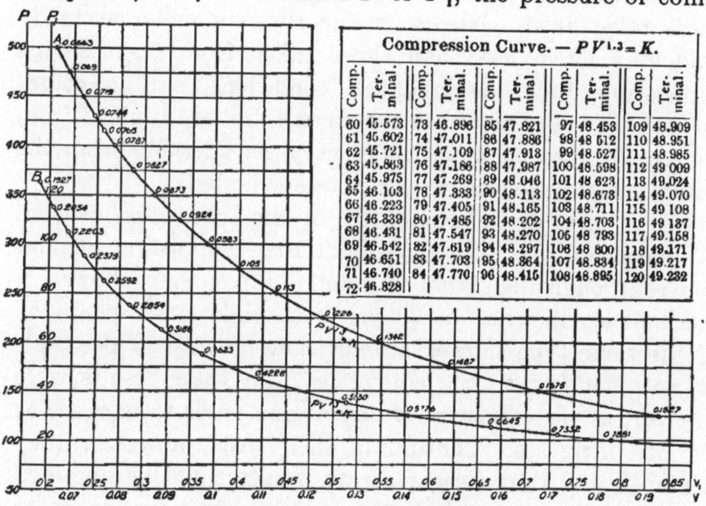


Compression Curve. — $PV^{1.3} = K$.

| Comp. | Terminal. | Comp. | Terminal. | Comp. | Terminal. | Comp. | Terminal. | Comp. | Terminal. |
|---|---|---|---|---|---|---|---|---|---|
| 60 | 45.573 | 73 | 46.896 | 85 | 47.821 | 97 | 48.453 | 109 | 48.909 |
| 61 | 45.602 | 74 | 47.011 | 86 | 47.886 | 98 | 48 512 | 110 | 48.951 |
| 62 | 45.721 | 75 | 47.109 | 87 | 47.918 | 99 | 48.527 | 111 | 48.985 |
| 63 | 45.863 | 76 | 47.186 | 88 | 47.987 | 100 | 48.598 | 112 | 49 009 |
| 64 | 45.975 | 77 | 47.269 | 89 | 48.046 | 101 | 48 623 | 113 | 49.024 |
| 65 | 46.103 | 78 | 47.333 | 90 | 48.113 | 102 | 48.673 | 114 | 49.070 |
| 66 | 46.223 | 79 | 47.405 | 91 | 48.165 | 103 | 48.711 | 115 | 49 108 |
| 67 | 46.339 | 80 | 47.485 | 92 | 48.202 | 104 | 48.703 | 116 | 49 187 |
| 68 | 46.481 | 81 | 47.547 | 93 | 48.270 | 105 | 48 793 | 117 | 49.158 |
| 69 | 46.542 | 82 | 47.619 | 94 | 48.297 | 106 | 48 800 | 118 | 49.171 |
| 70 | 46.651 | 83 | 47.703 | 95 | 48.364 | 107 | 48.834 | 119 | 49.217 |
| 71 | 46.740 | 84 | 47.770 | 96 | 48.415 | 108 | 48.895 | 120 | 49.232 |
| 72 | 46.828 | | | | | | | | |

Fig. 27. Compression Chart.

pression and trace the horizontal ordinate passing through this point to its intersection with the curve; from the point of intersection trace the vertical ordinate to the horizontal ordinate $V$ or $V_1$, where the per cent of compression space required for the compression in question may be read. Inverting the order of procedure will give the pressure when the per cent of compression space is known. In the columns to the right are given the terminal pressures for the expansion curve. With the curve and the values of the terminal pressures given, in order to construct a complete ideal card it is only necessary to compute the intermediate points for the expansion curve.*

* In using the curves to obtain values from 500 down to 125 lb., use the reference ordinates $P$ and $V$ in connection with the curve $A$. For values from 125 down to 20 lb., use the reference ordinates $P_1$ and $V_1$ in connection with curve $B$.

As stated in the chapter on "Engine Testing," the mechanical efficiency of any engine is the ratio of the b. hp. to the i. hp. It is the general supposition that the i. hp. as obtained from an actual card is correct within reasonable limits. Unfortunately experience has proven this to be erroneous in tests where proper care was not exercised. The cards taken from a steam engine running under a steady load vary much less than those taken from a gas engine supposedly under as favorable conditions, and in consequence, it is apparent that the mean i. hp. as obtained from the cards taken in a test may vary considerably from the actual mean power developed in the cylinder.

In the tests of the committee of the Institution of Civil Engineers on the efficiency of internal-combustion engines it was brought out that a small 5-hp. engine showed a mechanical efficiency of 90 per cent, while a 20-hp. engine of the same type showed an efficiency of only 80 per cent. It was manifest to the committee that these results were in error, as the efficiency of any engine tends to increase as its size increases. Continuing their tests, however, they demonstrated that very accurate results could be obtained if a properly designed indicator was used and care taken to maintain a uniform gas pressure. A paper by Bertram Hopkinson, read before the Institution of Mechanical Engineers and excerpts of which may be found in *The Engineering Magazine* for January 1908, gives the results of his extensive experiments along these lines and shows that he was able to obtain diagrams correct to within 1 or 2 per cent. It may be added that the pressure of the cooling water should also be maintained as nearly uniform as possible, in order to secure the best results.

# CHAPTER XII.

## GENERAL DIMENSIONS.

THE general dimensions discussed in this chapter may be taken to mean the fundamental dimensions, the use of which is necessary to produce a certain desired power condition in the engine. The power of any engine, either of the internal or external combustion type, is dependent upon four conditions — the mean effective pressure acting on the piston during the working stroke, the length of stroke or duration of this applied pressure, the area of the piston, and the number of times per minute the pressure is exerted. As stated in the chapter on "The Indicator Card," the power thus computed is termed the theoretical horsepower, while the power obtained by test is designated the indicated horsepower (i. hp.). The delivered horsepower, or, as it is more commonly designated, due to the method used in its determination, the brake horsepower (b. hp.), is some smaller value, usually from 75 to 90 per cent of the i. hp. The mechanical efficiency is the ratio of the b. hp. to the i. hp., or $\frac{\text{b. hp.}}{\text{i. hp.}}$, the ratio being expressed in terms of a per cent. The friction horsepower is equal to the difference between the i. hp. and the b. hp., or friction hp. = i. hp. − b. hp.

Since different fuels have varying heat values, the force of explosion, as already explained, varies with the fuel and is directly proportional to the cylinder compression before explosion. Hence the mean effective pressure in any cylinder is dependent on two general factors, the quality of the fuel and the cylinder compression. The proportions of the explosive mixture naturally affect the force of the explosion, but since, in the computation of the power to be delivered by any cylinder, the mixture is assumed to be theoretically correct, this factor need not be considered in the design. The highest mean effective pressure is obtainable by the use of gasoline, with natural gas a close second; then come the illuminating gases in the order of their candlepowers, then water gas, producer gas, and the poorer quality of gas derived from the blast furnaces.

The average mean effective pressures range from 40 lb. per sq. in. to 97 or 98 lb. per sq. in., the latter values being found in the Diesel oil engines. An average value for natural gas or gasoline is from 65 to 70 lb. per sq. in. The above values are all gage pressure readings.

The cylinder compression having been determined upon and the theoretical card constructed, as described, the mean effective pressure obtained from this card may now be used to determine the bore and stroke necessary to produce a given b. hp.

Let $d$ = diameter of cylinder in inches.
$l$ = length of stroke in inches.
$n$ = revolutions per minute.
$p$ = mean effective pressure.
$P_b$ = brake horsepower.

In this formula the fuel used is supposed to be equal in heat value to natural gas or gasoline and the pressure of the explosion to be four times the compression pressure. For poorer gases a fuel factor must be introduced in the formula. Then for a four-cycle engine

$$P_b = \frac{d^2pln}{800,000} \qquad (6)$$

And for a two-cycle engine

$$P_b = \frac{d^2pln}{600,000} \qquad (7)$$

For an engine operating on producer gas and with the same compression the fuel factor may be taken as approximately 0.7. Producer gas engines, designed particularly for that purpose, generally carry their compressions much higher. (See chapter on "Compression.") For engines operating on blast-furnace gas a fuel factor of 0.5 to 0.6 is necessary.

The above formulas may now be used for determining the dimensions of any engine of like compression to be designed. For example, supposing it is desirable to build a single-cylinder, four-cycle engine capable of delivering 40 hp. and that the mean effective pressure has been determined at 70 lb. per sq. in. It is necessary, in order to solve the equation, that but one value be unknown. In order to accomplish this result we must assume the value for two of the unknown quantities and solve the equation to determine if the remaining unknown value is in the proper proportion. As the speed at which the engine is expected to run is generally known, we assume this value and the length of stroke, leaving the formula to be solved for *d*. In this case we will assume the engine speed to be 200 rev. per min. and the length of stroke to be 18 in. which would result in a piston speed of 600 ft. per min., approximately the correct value for a stationary, low-speed, horizontal engine.

Then
$$40 = \frac{d^2 \times 70 \times 18 \times 200}{800,000}$$

or
$$d^2 = 158$$

and
$$d = 12.6.$$

The percentage of compression space of the entire cylinder volume being known, it becomes an easy matter to determine the length of the space, having determined the value of *d*. For example, supposing that the per cent of compression space necessary to obtain a mean effective pressure of 65 lb. was found to be 23, or the compression space was 0.23 of the total cylinder volume.

Then $V = v + 0.23\ V.$

In which $V$ = the total cylinder volume and
$v$ = piston displacement.

Solving we obtain the value of 0.23 $V$, and, the area of the compression space being known, we are able to determine its length.

The proper ratio of bore to length of stroke is, for stationary four-cycle engines, as 1:1.3 or 1.5. For two-cycle engines this ratio should be decreased, some engines even being made square, that is, with the stroke equal to the bore. Automobile and marine engines, owing to the requirement that they set low, are made with their stroke from 1 to 1.2 times the bore, and often run at a piston speed as high as 800 ft. per min.

In the design of an engine it is often found expedient to follow as closely as possible the general ratio of dimensions of some engine of the same type, the operation of which has been found by experience to be uniform and satisfactory. As an example take an engine with a cylinder diameter of 12 in., a stroke of 18 in., and a mean effective pressure as shown by the card of 60 lb. per sq. in., which has been found to give good satisfaction at 190 rev. per min. and to deliver 30 b. hp. when operating on natural gas as a fuel.

Then applying our formula for four-cycle engines we have, as an unknown value, the value of the denominator of the fraction. Calling this value $x$ the formula may be written as follows:

$$30 = \frac{12^2 \times 60 \times 18 \times 190}{x}.$$

Solving for $x$ we find its value to be 984,960.

Then the equation for a similar engine of any desired horsepower would be

$$\text{b. hp.} = \frac{d^2 pln}{984{,}960}.$$

In a similar way a formula may be found for any engine the operation of which is known.

Formulas 6 and 7 may be simplified and made to apply to special cases only. That is, when the piston speed is determined as 600 or 700 ft. per min., the diameter of the cylinder may be found by means of the following simple formulas.

| Four Cycle. | | Two Cycle. | |
|---|---|---|---|
| Piston Speed. | Diam. of Cyl. | Piston Speed. | Diam. of Cyl. |
| 600 | $18.6\sqrt{\frac{p_b}{p}}$ | 600 | $22.3\sqrt{\frac{p_b}{p}}$ |
| 700 | $17.3\sqrt{\frac{p_b}{p}}$ | 700 | $20.1\sqrt{\frac{p_b}{p}}$ |

Similar formulas may be derived for any given piston speed.

Having found the value of $d$, the length of stroke may be determined by making its length in the proper proportion to the diameter, as previously described.

The computations described and the formulas given in this chapter are for a single cylinder only. In the design of a multiple-cylinder engine it should be borne in mind that the number of cylinders must govern the size of the individual units; thus the individual cylinders of a four-cylinder engine should be designed to develop slightly less than one-quarter as much power as a single-cylinder engine of the same horsepower. At first thought it may appear to the reader that the four cylinders should each have exactly one-quarter the power of the single large cylinder. This conclusion is in error owing to the fact that the mechanical efficiency of a multiple-cylinder engine is from 5 to 10 per cent greater than that of a single cylinder, owing to the better balance secured with the consequent diminution of the friction loss.

# CHAPTER XIII.

## THE CAM MECHANISM.

PROBABLY the most important mechanical feature in the design of an internal-combustion engine is the cam mechanism, as nearly every engine with mechanically controlled valves uses this mechanism as a basis for transforming the rotary motion of the crank or cam shaft into the straight-line motion necessary for the operation of the mushroom valve.

The cams in a four-cycle engine are commonly located on the cam or half-time shaft, so called because it is designed to revolve at a speed equal to half that of the crank shaft. The reason for this is apparent when we consider that in a complete Otto cycle the crank shaft makes two revolutions, and since the exhaust and inlet valves open but once during this cycle, their operating mechanism must work half as fast as the crank shaft.

Occasionally engines are manufactured which use some mechanical principle other than 2 : 1 reduction gearing and cam shaft to attain this end. An eccentric mechanism may be devised, or other suitable methods may present themselves to the designer. Fig. 28 represents a half-time mechanism operated by means cf an eccentric on the crank shaft. In many small engines, as has been previously mentioned, only the exhaust valve is mechanically operated, the inlet being of the suction type.

It is good practice, and one adhered to in the manufacture of most of the large engines, to transmit the cam motion to the valve stem by means of an intermediate member. The majority of the smaller engines of the automobile and marine types interpose a push rod between the valve stem and the cam — it is not good design to allow the valve stem even when carrying a good steel roller to take its motion directly from the cam, as the side thrust will tend to wear the rod bushing. If for the purpose of reducing cost this method is pursued, the stem must be given a good liberal bearing to minimize this tendency as much as possible. The cam should also be set out of line in the direction toward which the eccentric portion is rotating as it approaches the push-rod roller; this will reduce the component of the cam movement acting at right angles to the axis of the valve stem. In Fig. 29 *ab* is the center line of the valve stem and roller; *cd* is the center line of the cam; *A* is the guide block, of squared stock, forked at the lower end to carry the hardened steel roller, and having the push rod tapped into its upper end and locked with a thin lock nut, *C*, as shown, thus providing adjustment of the mechanism so that the cam may commence to lift the valve at the proper instant; *B* is the guide, usually of brass, although it may be integral with the case, D. Even when an intermediate lever or push rod is interposed, the offsetting of the cam is good practice.

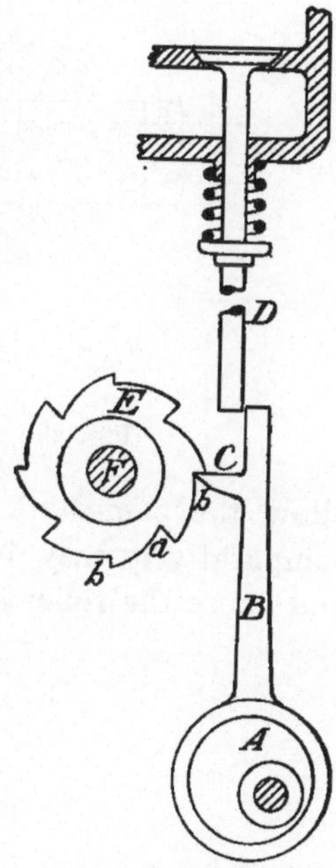

Fig. 28. Eccentric Half-time Mechanism.

A common form of cam mechanism with lever transmission device is shown in Fig. 30; *a* being the hardened steel roller, on which the cam operates, and *b* a hardened steel pin, or other suitable device, which impinges on the push rod *c*. Bearing *d* carries the transmission lever *e*. This mechanism is used quite extensively where the engine is started by compressed air, and at least one exhaust cam is made double. In order to bring this double-acting cam into starting position, it becomes necessary to shift either the cam or the cam roller, the latter method being the more convenient, owing to the fact that to shift the cam shaft would entail throwing the half-time gears partly out of mesh, or making the crank-shaft pinion with wide enough face to

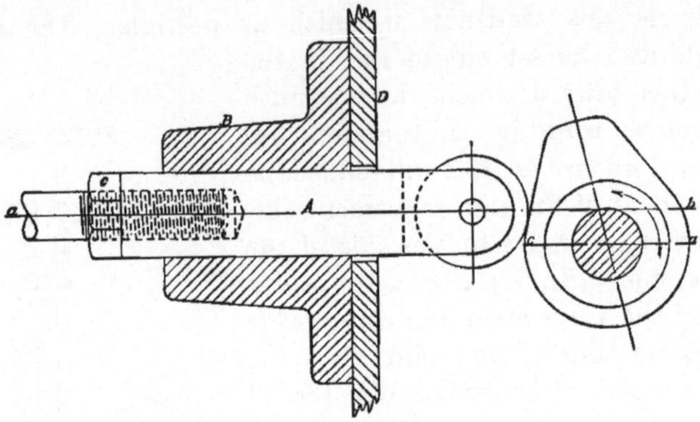

Fig. 29. Relation of Cam to Push Rod.

allow the cam-shaft gear its necessary movement. It is comparatively easy to make shaft *d* the shifting medium, and cause the roller *a* to move into the proper position for

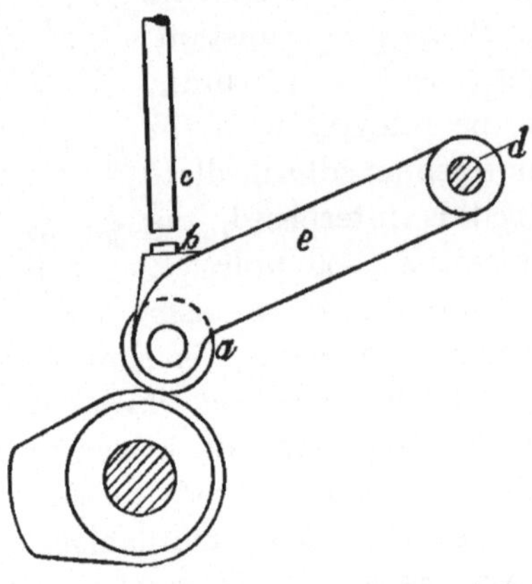


Fig. 30. Cam with Lever Transmission.

starting; pin *b* must be replaced either by two pins or a hardened steel block of sufficient length to allow the lever *e* to move into position for starting and yet maintain a contact between *b* and *c*. The method of starting by compressed air will be described in the chapter on "Starters."

Cams are of two classes: — (1) Spring, or gravity, return cams. (2) Positive return, or yoke cams. The style in universal use in gas-engine practice is the spring return system.

In laying out a cam, the method as diagramed in Fig. 31 will produce accurate results and, with slight modifications, will be found suitable for nearly every case. From the chapter on "Valves" we obtain data from which we are able

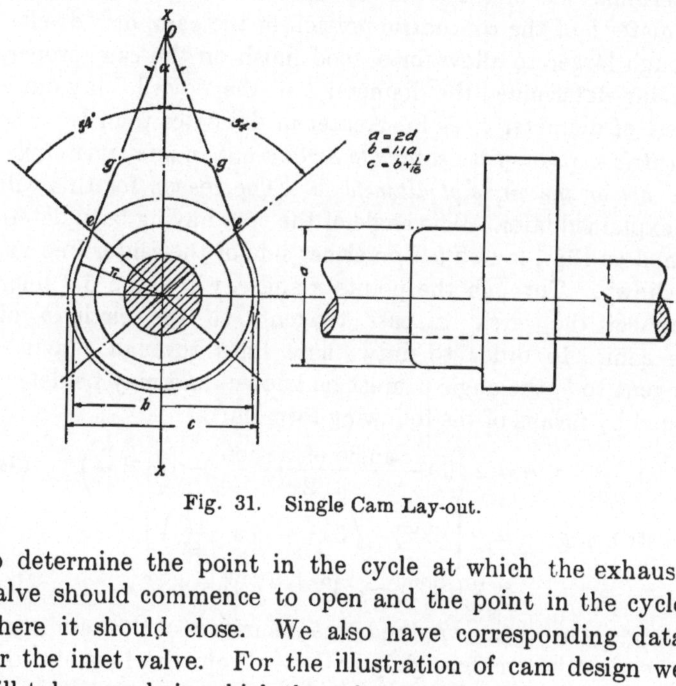


Fig. 31. Single Cam Lay-out.

to determine the point in the cycle at which the exhaust valve should commence to open and the point in the cycle where it should close. We also have corresponding data for the inlet valve. For the illustration of cam design we will take a cycle in which the exhaust valve is to open when the piston has completed 0.85 of the working stroke, and to close when the piston has completed 0.05 of the suction stroke. The cam will then consist of two parts, a portion *ee*, concentric to the shaft, and the eccentric portion *egge*, which eccentric portion is to be determined from the location of the two points above mentioned and the lift of the valve required. The period of opening of the exhaust valve is in terms of the crank motion, $1.2 \times 180° = 216°$ (since the valve opens 0.15 before the end of the explosion stroke and closes 0.05 after the beginning of the suction stroke). On the cam shaft, then, the period would be equal to $\frac{216}{2} = 108°$.

Having determined the period of opening on the cam shaft, proceed as follows: the size of the cam shaft being known, determine the diameter to be used for $a$, and make the diameter $b$ of the concentric portion of the cam, or "dwell," enough larger to allow for a good finish on the cam proper. Having determined the diameter $b$ of the "dwell," lay out a circle of diameter $c$, $\frac{1}{16}$ in. greater in diameter than $b$. *The eccentric portion of the cam is to be laid out on this latter circle, and not on the circle of diameter b.* The reason for this will be explained later. The angle of the cam having been determined as 108°, lay off 54° on either side of the center line $xx$, as shown. Through the points $e$ and $e$, where the 54° lines intersect the circle $c$, pass tangents to the circle $b$ of the cam. In order to draw these lines through $e$ and $e$ tangent to $b$, the angle $\alpha$ must be known, and may be determined by means of the following formula:

$$\alpha = 2\left(90 - \frac{\text{angle of action}}{2} - \cos^{-1}\frac{r_1}{r}\right)^* \qquad (8)$$

Substituting: $\alpha = 2\left[(90°) - \left(54° + \cos^{-1}\frac{\frac{11}{16}}{\frac{23}{32}}\right)\right]$

$$= 2\,[(90°) - (54° + 17°)] = 38°$$

Angle $\alpha$ being known, the tangent surfaces of the cam may be readily drawn through points $e$. From the formula for

* In the above trigonometrical formula, the term, $\cos^{-1}\frac{r_1}{r}$, means the angle whose cosine is $\frac{r_1}{r}$. In the cam in question $r_1 = \frac{11}{16}$ and $r = \frac{23}{32}$. In decimals, $r_1 = 0.6875$, $r = 0.7187$, and $\frac{r_1}{r} = 0.9565$. From the table of functions we find that the angle whose cosine is 0.9565 is about 17°. (It is really 16° 57′.)

effective valve areas we have determined on the lift necessary for the valve which this cam is to actuate. Supposing the lift to have been determined as 0.25 in.; lay off a concentric circle of 0.25 in. greater diameter than circle *c* and, when this circle intersects lines *Oe*, the cam outline will be completed. Corners *g* should be slightly round in the machining, so as to make the movement of roller from the tangent rise to the "dwell" *gg* as gradual as possible, without materially affecting the valve opening.

The reason for laying out the eccentric portion of the cam on a circle $\frac{1}{16}$ in. larger in diameter than the cam itself is that there may be no lost motion to take up during the angle of action. As in most cam mechanisms there is a slight clearance allowed between the valve stem and the member interposed between it and the cam, this clearance must be taken up before the valve can possibly commence to open. Added to this clearance allowed there are almost sure to be some other slight losses, due to looseness of bearing pins or to wear of parts. (Especially is this true in old engines.) Now when the roller traveling on the cam circle *b* encounters the tangent portion of the cam it commences to rise, and when the roller passes the point of intersection of this tangent and the circle *c*, on which the angle of action is laid out, if our assumption as to the amount of lost motion is correct, the valve should be just commencing to open, and should remain open until the roller passes the intersection of the other tangent surface and circle *c*. As a matter of fact $\frac{1}{32}$ in. is probably a little too much to allow for play, and may be reduced, if the designer sees fit, to a smaller amount; but it is doubtful if, after the engine has been run a few months, $\frac{1}{32}$ in. would be at all excessive.

Having laid out the exhaust cam, the inlet cam may be drawn in the same manner, the angle of operation of the cam being determined in the same way as for the exhaust.

Some manufacturers obtain the opening of the exhaust and inlet valves by means of a double cam in connection with a rocker arm. Fig. 32 gives the general idea of this cam and valve mechanism. Rocker arm *a*, pivoted at *b*, is provided with two bearing points, *c* and *d*, which operate, respectively, the exhaust and inlet valves, *e* and *f*. The push rod is connected at *g*, as shown, and its movement is restricted by the coil spring *h*. The push-rod mechanism may be identical with that used for an ordinary simple cam,

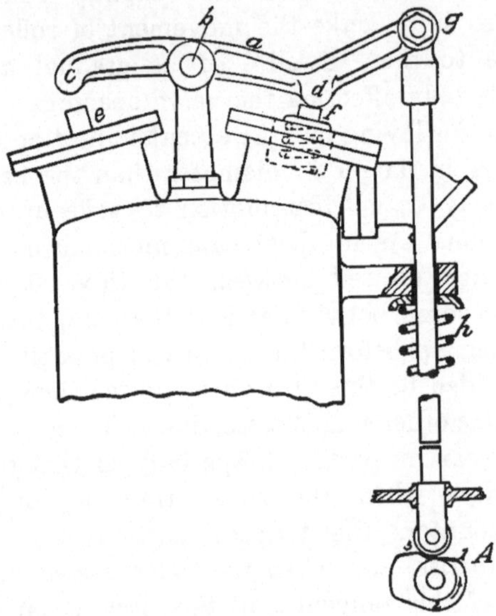


Fig. 32. Double Cam.

and need not be described. The joint at *g* must be made compensating, to prevent binding, as its center follows the curve *xx*. Fig. 33, showing a sectional view of a vertical engine, illustrates the method of double cam valve mechanism as it has been applied to automobile engines. The general outline of cam is shown at *A*, Fig. 32, but in practice the cam outline, instead of dropping suddenly into the depression as shown, is made to cut across on a chord of the cam circle, as shown in Fig. 33. While this reduces the noise of operation, it plainly reduces the mean effective opening, as will be later described. The operation of the double cam is as follows: in the position shown, the inlet valve *f* is wide open, the spring *h* having caused the lever *a* to move about *b* and to act on the stem of *f*. The cam roller, in the meantime, follows the outline of *A*. When point 1 is reached the

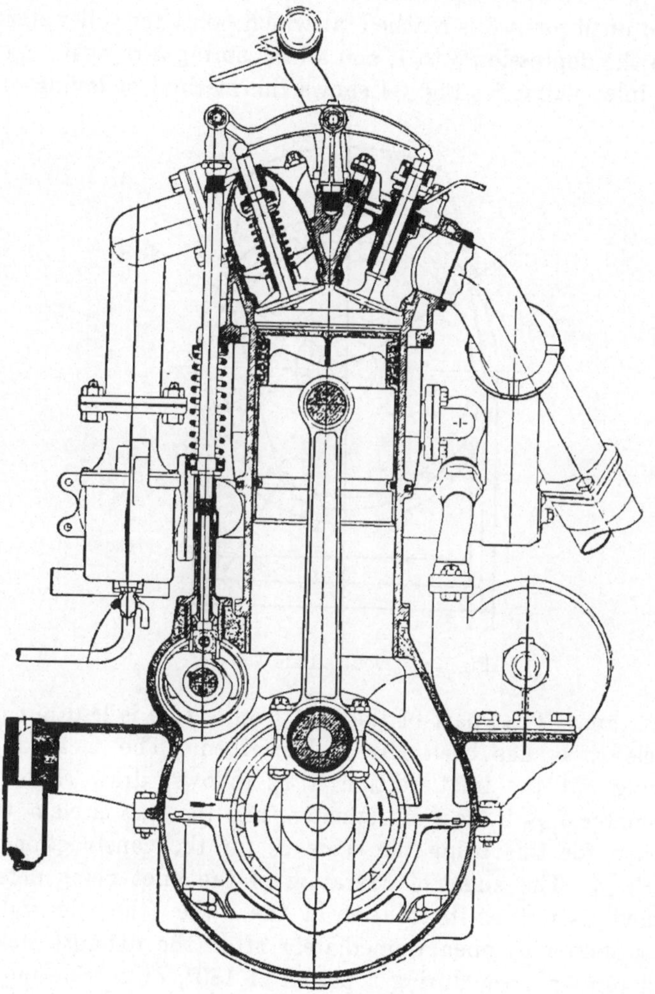

Fig. 33. Section of Vertical Motor Mechanism.

push rod is raised, until valve *f* is completely closed. During the "dwell," from 1 to 2, both valves remain closed, with roller *c* about to engage the stem of valve *e*, and just out of contact with the stem of *f*. When the roller reaches the tangent portion of the cam, at 2, the push rod rotates lever *a* about *b*, causing the exhaust valve *e* to open and remain open until point 3 is reached, at which point the roller starts into the depression 3 to 1, and allows spring *h* to again open the inlet valve *f*. Fig. 34 shows the method of laying out

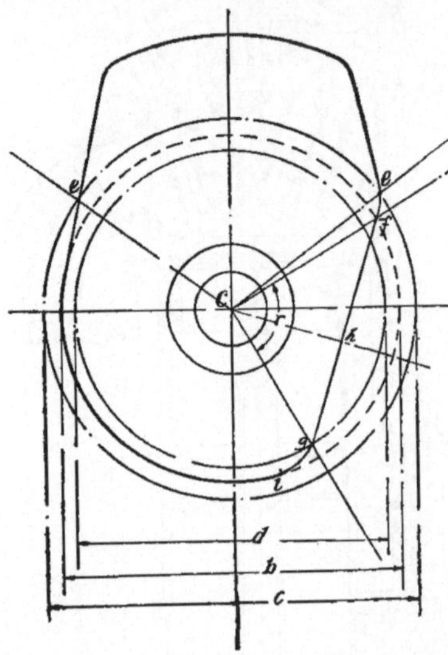

Fig. 34. Double Cam Lay-out.

this cam. That part of the cam from *e* to *e* is laid out on circle *c*, as has been already described. The method of laying out the inlet portion is as follows: draw circle of diameter *d* $\frac{1}{16}$ in. less in diameter than the cam circle *b*, the reason for this being the same as for the construction of circle *c*. The angle of operation of the inlet being determined, as well as its location in the cycle (the inlet valve is supposed to open immediately after the exhaust closes and remain open during a period of 180°, or a trifle more, on the crank circle. In this particular cam we will keep the inlet valve open through exactly 180°, and cause it to open immediately after the exhaust closes), lay off angle $\gamma$, as shown, equal to 90°. It is obviously impossible to lay off the inlet portion from *Ce* without cutting down the effective opening of the exhaust valve in order to obtain the curve from *e* to *f*. In order to overcome this difficulty, lay off *Cf* four or five degrees from *Ce*, and lay off *Cg* 90° from *Cf*. Draw the curves *ef* and *gi*, as shown, tangent, respectively, to the tangent rise and the "dwell" of the cam, and connect the points *f* and *g*. A little study of the cam outline will show that the period of maximum opening of the inlet valve is but an instant, the valve opening to its maximum at *h*, and immediately commencing to close again. Due allowance must be made for this feature, and the valve made large enough to produce an average opening of sufficient area to prevent throttling the charge. While the double cam reduces the number of parts necessary, it is claimed, and with good reason, to be more noisy than two separate cams.

The cam is usually made of machine steel, case hardened, as are also the rollers. They should be located on the cam shaft and secured in place with taper pins. The angular location of the exhaust and inlet cam, if on the same shaft, is an important operation, and, unless great care is used, the timing of the valves will be wrong. Either cam may be readily located alone, but to secure their relative locations being correct requires extreme precision. The best method to use, if manufacturing in quantities, is to provide a drilling and reaming fixture for the shaft and cams in which the relative position of the two cams has been accurately located, and in which position they are held while the pin holes are being made. The gear may then be keyed to the shaft and the cams set to their proper location in the cycle.

### The Reduction Gearing.

The gears connecting the cam shaft to the crank shaft are of the ordinary spur variety if the cam shaft and crank shaft are parallel; if the cam shaft is at an angle to the crank shaft, but in the same plane, bevel gears are used; if at an angle, and not in the same plane, skew gears must be used. The spur gear consists, essentially, of a disk, the teeth of the gear being cut on its periphery; the bevel gear has its teeth cut on the surface of a cone. The skew gear has its teeth cut on the periphery of a disk, but at such an angle with the shaft that they will mesh with the teeth of a gear the shaft of which is neither parallel nor in the same plane. Fig. 35 shows these three types of gears; at *a* is shown a pair of spur gears, the gear marked 1 being the crank shaft pinion and the one marked 2 the cam shaft gear. The size of pinion 1 is half that of gear 2, in order to secure the two to one speed reduction required. At *b* is shown a pair of bevel gears for

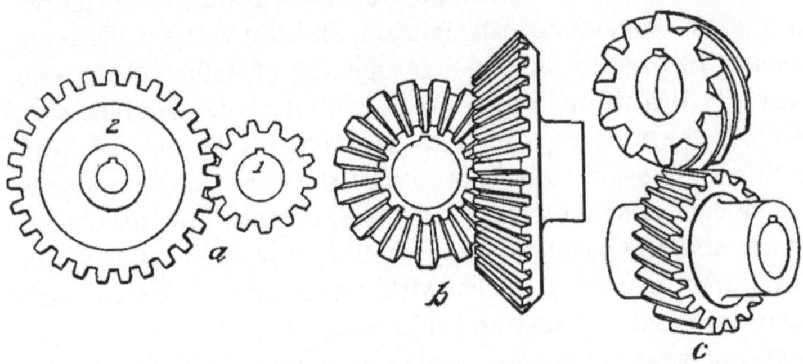


Fig. 35. Types of Gears.

changing the direction of motion and at the same time making the two to one reduction. The center line of the bevel pinion 3, which is carried on the crank shaft, and that of the bevel gear 4, which is the cam shaft gear, lie in the same plane and, in this particular example, at right angles. As in the use of the spur gears, the size of the crank shaft pinion is made half that of the cam shaft gear, in order to secure the proper speed reduction. At *c* is shown a pair of skew gears, by the use of which the speed reduction may be secured and at the same time maintain the crank shaft gear as large or, as most frequently is the case, larger than the cam shaft gear. The skew affords a quiet method of drive but one in which the gears, due to their action, are subject to an abnormal amount of wear. As the design of gearing is a subject by itself, and one on which several comprehensive works have been written, the author will not consume valuable space in it details. Cut gears are usually bought direct from manufacturers or, if the output of a concern warrants it, a gear-cutting machine is installed. In either case it is unnecessary to give the maker more than the pitch diameter, diametral pitch, and width of face of any spur or bevel gear to insure the teeth being absolutely accurate. For a pair of skew gears give the speed ratio, the number of teeth, the face, and the pitch and pitch diameter. In such a pair of gears $v : V :: D \cos A : d \cos a$.

In which
$v$ = velocity of smaller gear.
$V$ = velocity of larger gear.
$d$ = pitch diameter of smaller gear.
D = pitch diameter of larger gear.
$A$ = spiral angle of larger gear.
$a$ = spiral angle of smaller gear.

In locating the cam and crank shaft centers care must be taken to keep them such a distance apart that the cam shaft gear will mesh with that of the crank shaft and leave $\frac{1}{32}$ in. clearance between their pitch circles. For instance, supposing the pitch diameter of the crank shaft pinion were 4 in. and that of the cam shaft gear 8 in., then the distance between the crank shaft and cam shaft center lines should be $\frac{4+8}{2} + \frac{1}{32} = 6\frac{1}{32}$ in.

Whether the spur, bevel, or skew gearing is used, the pitch should be made small in order to secure quiet running and a minimum amount of back lash, as well as to make it possible to change the valve timing. On small engines 10 or 12 pitch is the rule, but on larger engines and on those for general rough use this may be increased to 7 or 8. For finer adjustment of the valves than may be secured by shifting the gear a tooth at a time, the cam shaft gear is sometimes made with the rim adjustable, so that the cam may be placed in any desired position in relation to the crank shaft. Fig. 36 shows such an adjustable gear. In the figure, $a$ is the central disk, on which the ring carrying the teeth is placed, as shown, joint $cd$ being a close machined fit. The outer ring is secured to the disk by bolts $e$, as shown, and adjustment is secured by means of the slots $f$. To further insure quiet running, the cam shaft gear is often made of fiber, or rawhide, and brass. The fiber or rawhide is held between two brass plates and the three are bolted or riveted together

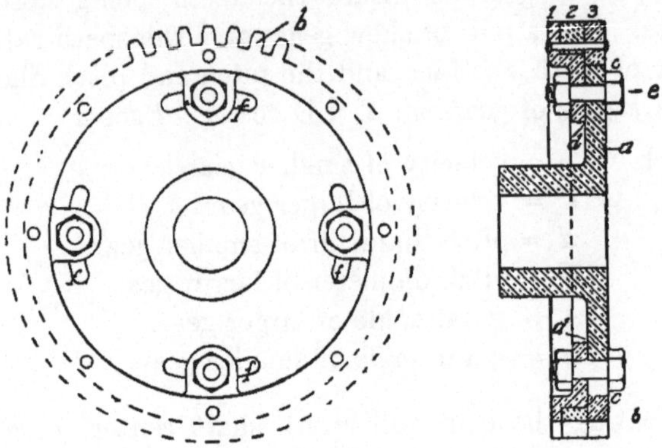


Fig. 36. Adjustable Spur Gear.

and then cut as one piece. The gear shown in Fig. 36 is constructed in this way, parts of the rim marked 1, 2, and 3 being brass, fiber, and brass, respectively.

# CHAPTER XIV.

## THE VALVES AND PORTS.

THE valves of the standard gas engine are of the mushroom type, and may be flat or angle seated — most valves being of the latter variety. See Fig. 37 for illustrations of these two kinds of mushroom valves.

The flat seated valve is most available for use as the suction valve in engines using this type of inlet, as the comparatively

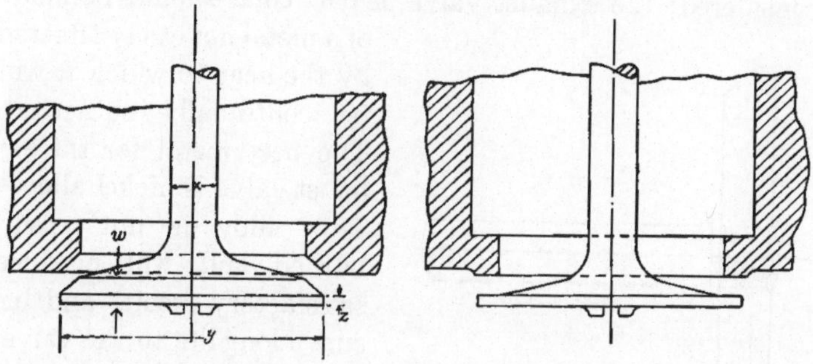


Fig. 37. Types of Mushroom Valves.

light weight, compared to an angle seated valve, is desirable, since lightness of the moving parts is an important feature in order that the inertia may be reduced to a minimum amount.

The angle seated valve for all purposes other than the one mentioned is far superior. The valve, meeting its seat at an angle, tends to wedge itself into place and seat more firmly; there is less tendency for carbon to collect and a greater tendency for the valve to grind out what may collect. The flat seated valve, being lighter, has greater tendency to warp and leak.

Valves should be drop-forged or cast in one piece, and then machined to finish size, as any two-piece valve tends to pull apart, whether the stem be tapped or riveted into the head. If tapped or riveted, the head must, of necessity, be made heavier and clumsier in order to give sufficient body for the stem. Fig. 38 illustrates a two-piece valve tapped and riveted.

The material of which the valve is to be made should be considered; the exhaust valve in particular should be made of a metal not easily affected by the heat to which it will be continually subjected. The best metal for the exhaust valve is nickel alloy.* This alloy is not easily burned and will produce satisfactory results and insure a long life to the valve. The inlet valve, while not subjected, except on its surface, to the hot exhaust gases, should, nevertheless, be made of a tough steel. Many manufacturers use nickel steel for the inlet, with excellent results.

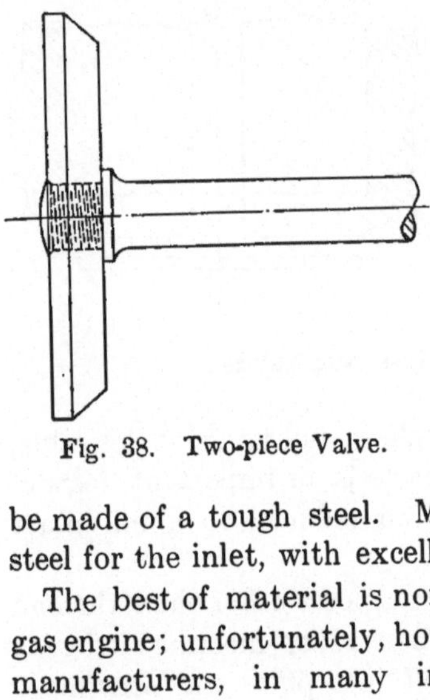

Fig. 38. Two-piece Valve.

The best of material is none too good for the valves of a gas engine; unfortunately, however, this does not prevent the manufacturers, in many instances, from using inferior material, with more or less successful results. Many engines have both the inlet and exhaust valves made of ordinary drop-forged steel stock, but while these valves may, in some cases, give excellent results for a time, their use in the manufacture of strictly high-class engines should be discouraged.

The size of the effective valve openings and of the inlet

* So-called nickel alloy is not the German silver of commerce. Alloy suitable for valves may be manufactured, or purchased direct from the manufacturers, who use their own formula. A manganese, copper alloy, containing 1.25 per cent of aluminum, is also a tough non-corroding material.

and exhaust passages is, or should be, based entirely on the probable speed which the gases will attain in passing through them: 6000 ft. per min. may be safely assumed as the speed of the inlet gases, and 5000 ft. per min. as the speed in the exhaust. The assumption is that the exhaust gases are at atmospheric pressure, but, as a matter of fact, their pressure will be more nearly in the neighborhood of 30 lb. per sq. in. The speed of 5000 ft. per min. is then a safe and conservative estimate.

It is an easy matter to design the passages large enough to accommodate these speeds, but the port or valve opening is liable to be restricted unless especial care is used in proportioning it. The following formulas will give a short method of determining the size of the passages.

Let $L$ = Length of piston stroke, in in.

$R$ = Rev. per min. of the crank shaft.

$S$ = Piston speed, ft. per min. $= \frac{2\,RL}{12}$

$N$ = Number of cycles per min. $= \frac{R}{2}$.

$D$ = Diameter of cylinder, in in.

$a$ = Area of inlet passage, in sq. in.

$a_1$ = Area exhaust passage, in sq. in.

$A$ = Area of piston, in sq. in.

$d$ = Diameter of inlet passage.

$d_1$ = Diameter of exhaust passage.

$$\text{Then } a = A\,\frac{S}{6000} \qquad (9)$$

It is apparent that if the piston speed were equal to the gas speed, $a$ would, of necessity, be as large as $A$.

Substituting for $S$

$$a = \frac{2\,LAR}{6000 \times 12}$$

$$= \frac{LAR}{36{,}000} \qquad (10)$$

And

$$a_1 = A\frac{S}{5000} \quad . \quad . \quad . \quad . \quad . \quad . \quad (11)$$

$$= \frac{2\ LAR}{5000 \times 12}$$

$$= \frac{LAR}{30{,}000} \quad . \quad . \quad . \quad . \quad . \quad . \quad (12)$$

In case the inlet and exhaust passages are circular in section, we may simplify the formulas in order to obtain $d$ and $d_1$ directly.

$$a = \frac{\pi d^2}{4}.$$

Then $$d^2 = \frac{4a}{\pi}$$

or $$d = \sqrt{\frac{4a}{\pi}}.$$

Also $$d_1 = \sqrt{\frac{4a_1}{\pi}}.$$

Substituting $$d = 2\sqrt{\frac{AS}{6000\ \pi}}$$

$$= 0.00527\ D\sqrt{RL} \quad . \quad . \quad . \quad (13)$$

And $$d = 0.00577\ D\sqrt{RL} \quad . \quad . \quad . \quad (14)$$

If the engine being designed is of the stationary type, for which 600 ft. per min. would be a good value of $S$, then we may use the following values:

$$a = 0.1\ A,$$
$$a_1 = 0.12\ A,$$
$$d = 0.316\ D,$$
$$d_1 = 0.346\ D.$$

These last named values are clearly special cases, covering but one class of design, and should not be taken as general values.

Table IX gives briefly the values of $a$, $q_1$, $d$ and $d_1$ for piston speeds of from 500 to 1200 ft. per min.

TABLE IX.

| Piston Speed. | $a$. | $a_1$. | $d$. | $d_1$. |
|---|---|---|---|---|
| 500 | .083 $A$ | .1 $A$ | .288 $D$ | .316 $D$ |
| 600 | .1 $A$ | .12 $A$ | .316 $D$ | .346 $D$ |
| 700 | .116 $A$ | .14 $A$ | .340 $D$ | .374 $D$ |
| 800 | .133 $A$ | .16 $A$ | .364 $D$ | .400 $D$ |
| 900 | .150 $A$ | .18 $A$ | .387 $D$ | .424 $D$ |
| 1000 | .166 $A$ | .20 $A$ | .407 $D$ | .447 $D$ |
| 1100 | .183 $A$ | .22 $A$ | .427 $D$ | .469 $D$ |
| 1200 | .2 $A$ | .24 $A$ | .447 $D$ | .489 $D$ |

The effective valve opening must next be considered. From Fig. 37, showing the angle and flat seated valves, it is apparent, after a little consideration, that in the former the effective opening of the valve is not the lift proper, but some

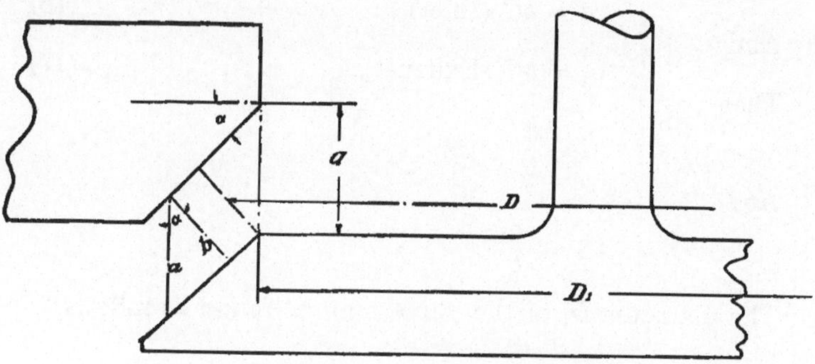


**Fig. 39. Diagram of Effective Valve Opening.**

function of it, depending upon the angle of the seat. Fig. 39, as a diagrammatic illustration, will make this fact clear. In the figure, the lift may be represented by $a$, but the effective opening is a smaller quantity, $b$. The angle of the seat may be made 45°, which is good practice, although the value is sometimes varied. Constructing the triangle as shown, we have

$$b = a \cos \alpha \quad . \quad . \quad . \quad . \quad . \quad . \quad (15)$$

In this case, the value of $\alpha$ being 45°, the formula becomes

$$b = a \cos 45°.$$

Considering $a$ as unity, the following table will give the values to be used for $b$ for angles of 40° 45°, 50°, 55°, and 60°.

| Angle. | a. | b. |
|---|---|---|
| 40° | 1 | 0.766 |
| 45° | ........ | 0.707 |
| 50° | ........ | 0.642 |
| 55° | ........ | 0.573 |
| 60° | ........ | 0.500 |

The effective opening is, of course, the annular space continuing entirely around the valve, and its area will be equal to the product of $b$ and the mean circumference $C$ of this annular space.

Then

$$a = bC \text{ (Inlet)} \quad . \quad . \quad . \quad . \quad . \quad . \quad . \quad . \quad . \qquad (16)$$

And

$$a_1 = bC \text{ (Exhaust)} \quad . \quad . \quad . \quad . \quad . \quad . \quad . \qquad (17)$$

Then

$$C = \frac{a}{b}$$

And

$$C_1 = \frac{a_1}{b}.$$

The diameter $D_1$ of the valve may be found as follows:

$$D = \frac{C}{\pi}.$$

Or, substituting

$$D = \frac{a}{\pi b} \text{ (Inlet)} \quad . \quad . \quad . \quad . \quad . \quad . \quad . \quad . \quad . \qquad (18)$$

And

$$D = \frac{a_1}{\pi b} \text{ (Exhaust)} \quad . \quad . \quad . \quad . \quad . \quad . \quad . \quad . \qquad (19)$$

Then

$$D_1 = D - b \sin \alpha$$

$$= \frac{a}{\pi b} - b \sin \alpha \text{ (Inlet)}$$

And

$$D_1 = \frac{a_1}{\pi b} - b \sin \alpha \text{ (Exhaust)}.$$

Simplifying,

$$D_1 = 0.316\frac{a}{b} - b \sin \alpha \text{ (Inlet)} \quad . \quad . \quad . \quad . \quad . \quad (20)$$

$$D_1 = 0.316\frac{a_1}{b} - b \sin \alpha \text{ (Exhaust)} \quad . \quad . \quad . \quad (21)$$

In the above formulas, the lift of the inlet and exhaust valves and the angle of their seats have been considered the same. Should it be advisable to increase the lift of the exhaust valve, rather than to increase its diameter, the value of $b$ in the exhaust-valve formula would, of course, be larger. Some judgment should be used in the application of the formulas, and in order to do so they should be thoroughly understood.

In the use of a flat seated valve, owing to the fact that the gas is deflected at a right angle in striking the valve and tends to produce swirls and choking, it is advisable to make the lift as great as for an angle seated valve of the same capacity, although the effective opening would be larger.

The size of valve stem, thickness of head, and other minor dimensions are details for which no fixed rule can be given. Materials as well as conditions will vary, and the best that can be done is to give sizes used in average practice and these, used with judgment, will give results not far wrong.

Referring to Fig. 37, we may make

$y = 1.18\ D_1.$

$x = 0.23\ D_1.$ Short stems.

$x = 0.28\ D_1.$ Long stems.

$z = 0.01\ D_1.$

$w = 0.09\ D_1.$

Valves may be set in the cylinder in a variety of different ways; designers agree, however, and common judgment tells us, that the best arrangement would be to place both the exhaust and inlet valves in the cylinder head and to do away with all pockets which, aside from their tendency to collect carbon and exhaust products, and to increase the radiating surface, weaken the pressure on the piston head. An ideal explosion chamber is a sphere, the pressure at all points on its interior surface being equal. Such an explosion chamber is impossible in a gas engine because the compression space would be altogether too large and the radiating surface of such extent as to render the idea impracticable. A modified form of such an explosion chamber has been tried by several manufacturers of small engines, the piston head being in the shape of a re-entrant cone and the cylinder head arched.

The setting of the valves in the cylinder head is, however, a common practice, but there are still many designers who adhere to the practice of setting the valves in pockets. Figs. 40 and 41 illustrate these two methods.

Fig. 40 shows the inlet and exhaust valve let into the cylinder head at an angle and designed to be operated by

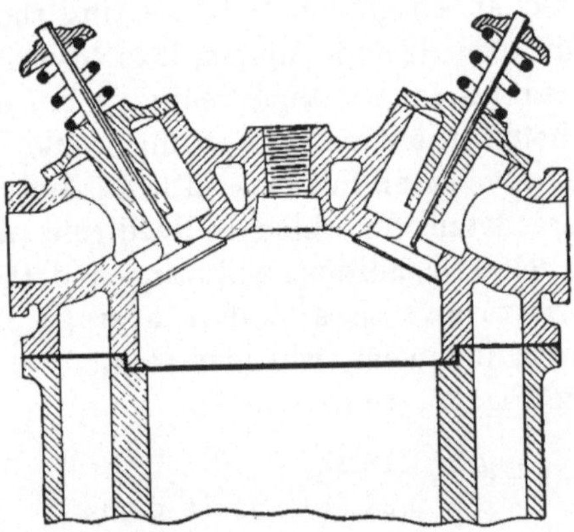

Fig. 40. Method of Setting Valves in Head.

means of a rocker arm and double-acting cam, as previously described in Chapter XIV. In this design no valve cages have been used, the cylinder head being cast separate and the valve seats ground into it as shown.

In the Rathbun gas engines both valves are let into the cylinder head in cages and operated by means of two separate cam mechanisms and levers.

In Fig. 41 the inlet valve is let in, in a cage, as shown. As there must be an opening $x$, greater in diameter than the exhaust valve head, in order that the valve may be put in place, the cage forms an easy and convenient method of bushing this opening and making both valves readily accessible. Details of the cage and cap are shown and need no further explanation. The cage and cap may both be of malleable iron. The cap yoke should be ribbed and strength-

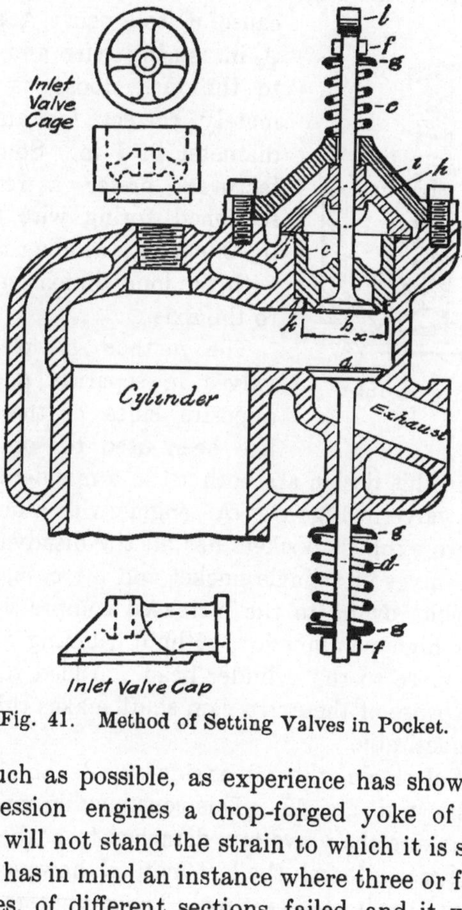


Fig. 41. Method of Setting Valves in Pocket.

ened as much as possible, as experience has shown that in high-compression engines a drop-forged yoke of ordinary dimensions will not stand the strain to which it is subjected. The author has in mind an instance where three or four drop-forged yokes, of different sections, failed, and it was found in this case that a yoke of manganese bronze gave the best satisfaction.

The method of inserting the inlet valve in a cage may be

used to good advantage in engines operating with the suction inlet. The only difference in design would be an increase in height of the cap in order to allow the valve spring to be placed in the cage. Fig. 42 illustrates such a valve assembled.

The spring for a suction inlet valve should be as light as possible, so that the slightest vacuum in the cylinder will cause it to open. A spring of $\frac{1}{16}$-in. spring wire and six coils to the inch, loose, is approximately correct for an outside diameter of $\frac{7}{8}$ in. Some manufacturers prefer a rectangular sectioned spring wire 0.042 in. by 0.068 in., winding the spring with the long dimension parallel to the axis.

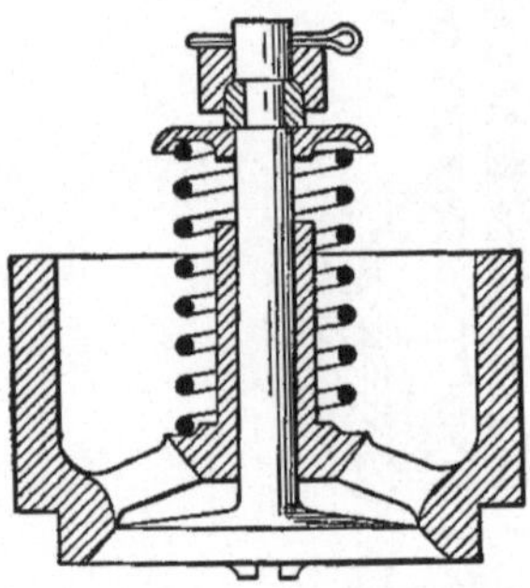

Fig. 42. Suction Inlet Valve in Cage.

The method of placing the valves in separate pockets, on opposite sides of the cylinder, has been used to some extent. The valves in this design are both set in a manner similar to the exhaust valve in Fig. 41. An engine with a double cam shaft and two exhaust pockets has all the disadvantages of placing the valves in a single pocket and more, inasmuch as it is impossible, owing to the increased compression space, to obtain a high compression without running the piston exceedingly close to the cylinder head. Added to this, the increased expense of the extra cam shaft makes this method of design undesirable.

A method of placing the valves side by side, and in a single pocket, is shown in Fig. 43. This construction may be used as shown, making the valve box detached from the cylinder, or it may be cast integral, the last method, however, making a much more complicated casting. The gas enters the valve box and cylinder through the inlet valve, as shown by the direction of the arrows *a*, and passes out of the cylinder through the exhaust as indicated by the arrows *b*.

Gas engines have been manufactured in which the valves were not of the mushroom type, the most notable of this class being the Otto slide-valve engine, which was used quite extensively at one time. The greatest difficulty experienced

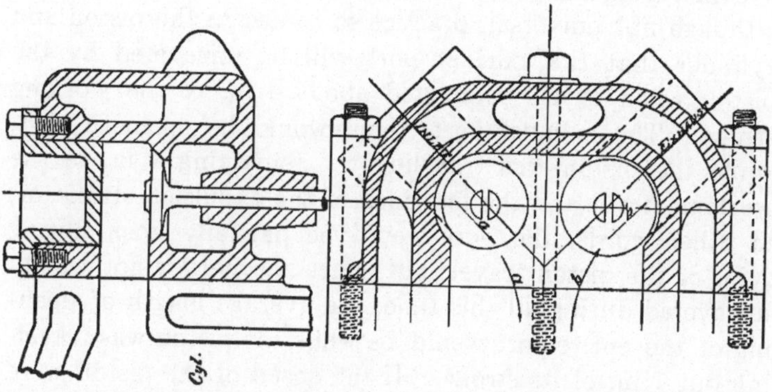

Fig. 43. Valve Pocket Cast Separate from Cylinder.

in the use of slide valves is the liability of their becoming hot and sticking, and, as the poppet-valve construction is cheaper than and as efficient as any other type, it naturally has become the most popular.

The two-cycle engine, if of the three-port type, makes use of no valves whatsoever; if the engine is of the two-port variety, but one poppet valve is used — between the carburettor and the crank case. This valve should be designed large enough to supply to the case a full charge during every suction stroke. The design of this valve is exactly similar to the cylinder inlet valve in four-cycle motors, and needs no further discussion.

The cylinder ports in a two-cycle motor must, however, be carefully proportioned, in order that, at any reasonable speed, the cylinder may receive a full charge of gas.

As previously explained under the head of two-cycle motors, the inlet and exhaust ports are open but a comparatively short space of time, whereas in the four-cycle type the inlet and exhaust ports are open during one complete stroke. It is obvious that, in order to secure a full charge, or a complete discharge of the exhaust products, the ports of a two-cycle engine must be proportionately much larger than the effective valve opening in a four-cycle engine of same size and designed to operate at the same speed.

Considering a two-cycle engine of 5-in. stroke, it is common, although not universal, practice so to design the piston and cylinder that the exhaust port will be uncovered by the piston when it has completed about 4 in. of the working stroke. The exhaust port, then, would be open, in part, while the piston was traveling the remaining 1 in. of the working stroke and the first inch of the expulsion stroke, or, in other words, the port would be partially open during 2 in. of the piston travel; but, since the port is not wholly uncovered during all this time, the average length of opening of the entire port would be while the piston was traveling but 1 in. of its stroke. If the speed of the piston were uniform during its complete stroke, then the period of total opening of the port would be approximately one-fifth of the total time required to complete one stroke, but the piston, during the period of opening of the port, is traveling at its minimum speed and in consequence the port is completely open during about one-fourth of the time required for a single stroke of the piston.

Since the port remains open but one-fourth as long as the valve of a four-cycle engine, it naturally follows that its effective area should be four times as large. It is a safe rule to follow to make the ports, both the exhaust and inlet, of a two-cycle engine four times the effective area of the valves of a four-cycle engine having the same bore, stroke, and piston speed.

The above rule is, of course, subject to some modification, and it is well to bear in mind that it is much better that the ports should be too large than too small, provided that, in order to increase the effective area, the exhaust port does not open too early in the working stroke, allowing the pressure in the cylinder to fall and materially weakening the power of the engine. The average two-cycle engine utilizes the expansive power of the gases during about four-fifths of the working stroke, the remaining one-fifth of the working stroke and the first one-fifth of the compression stroke being used to expel the products of combustion and to introduce the fresh charge.

The exhaust port is given a positive lead over the inlet port, in order that the pressure in the cylinder may fall and the products of combustion be partially expelled before the opening of the inlet port allows the rush of gas into the cylinder. If this were not done, unless the crank-case

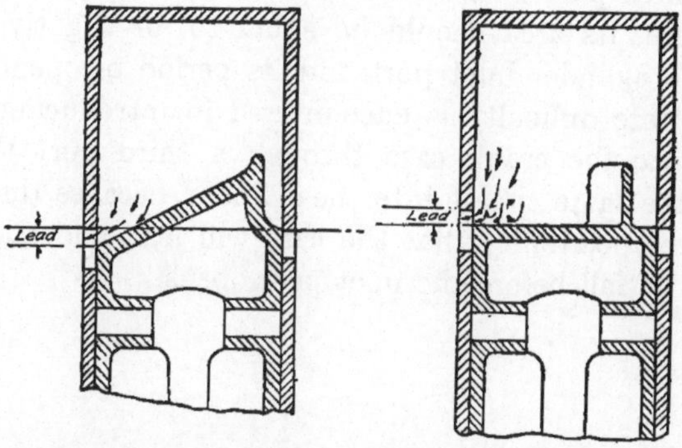


**Fig. 44. Methods of Obtaining Exhaust Port Lead in Two-Cycle Engines.**

compression were unusually high, the hot products of combustion would continually fire back into the case, due to their pressure being greater than the pressure in the crank case. The necessary lead is a point on which designers differ. However, if the exhaust port is made to open one-third to one-fourth of its width ahead of the inlet port, it will not be far wrong.

The lead of the exhaust is accomplished either by placing the port higher in the cylinder than the inlet or, as is becoming more and more the custom, by making the piston with the edge which uncovers the exhaust lower than the edge which uncovers the inlet, but making the ports even. The latter method is the better, since the inclined surface of the piston allows the hot gases to discharge out of the exhaust more readily, there being no sharp angle past which they must pass as is the case in the engines using a piston with a flat head. Fig. 44 illustrates these two methods clearly, the small arrows showing the paths of the exhaust gases.

As the ports continue for quite a distance around the cylinder, in order to prevent the piston rings slipping into them it is necessary that bars be cast across the opening at intervals, thus forming, instead of one large port, several smaller ports aggregating in area the required effective opening, the bars, forming a bearing surface, preventing the rings slipping into the openings and catching there.

If a third port is used for introducing the gas into the crank case, its area should be equal to, or slightly larger than, the cylinder inlet port and its period of opening the same. More difficulty is encountered in introducing a full charge into the crank case through a third port than in forcing the same charge into the cylinder, because there is a very great possibility that the case will leak and allow the vacuum to fall before the inlet port opens.

# CHAPTER XV.

## THE CYLINDER.

THERE are two classes of gas-engine cylinders, namely:

I. The air-cooled.

II. The water-cooled.

The air-cooled cylinder secures the cooling effect by radiation into the surrounding air. In order that this cooling effect may be as great as possible the radiating surface of the cylinder is made large by casting of annular rings or fins, about 0.25 in. apart and from 0.75 to 1 in. in height, from the top to the bottom of the cylinder and radiating from the center of the head, or by driving pins into holes drilled in the outside of the cylinder wall. The first named method is most commonly in use, although the pin method is claimed to have its advantages, allowing the air currents to circulate freely in every direction and reducing the first cost of the casting. However, the difference in cost of the castings would be more than made up in the drilling of the holes and setting the pins. Fig. 45 illustrates an air-cooled cylinder of the first class with valve cage and head cast integral.

Fig. 46 is one example of a water-cooled cylinder with the water jacket, head, and valve pocket cast integral. In this particular example the inlet valve is shown water-jacketed. This, however, is not necessary, as the cool gases rushing through this valve will keep it cool enough; however, in order to secure the very best cooling effect about the exhaust valve and outlet, it is advisable that the water jacket surround the valve pocket completely, thus making it possible to draw off the cooling water at *a* and thus maintain the outlet at the highest point of the water jacket, which condition is essential for perfect circulation. By drawing the water off at *a*, as shown, the valve box is thrown into the circulating system and the jacket water must all pass through its jacket to reach the outlet, whereas if the outlet *a* were placed on the top of the cylinder, the valve box would

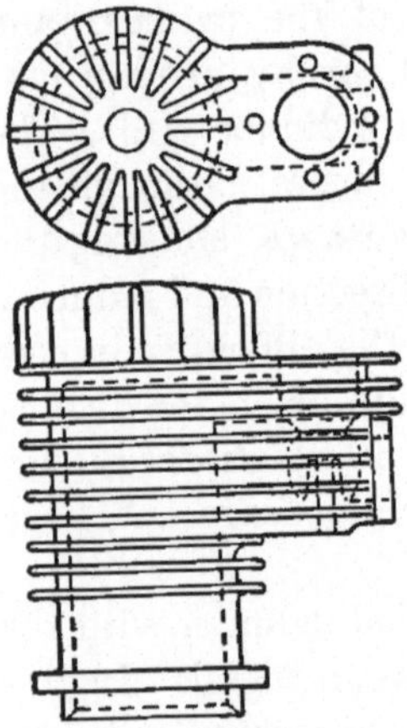

Fig. 45. Type of Air-cooled Cylinder.

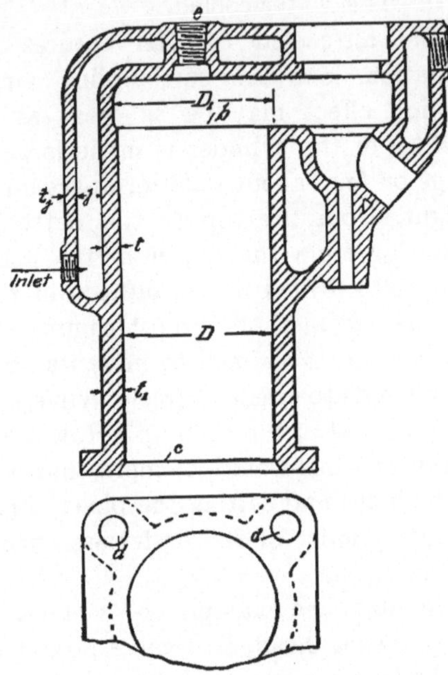


Fig. 46. Type of Water-cooled Cylinder.

be sidetracked, so to speak, and obtain but a very limited circulation.

The thickness of the cylinder walls to resist the internal pressure to which it is subjected may be found by the formula

$$t = \frac{pd}{2f} 0.1 \quad . \quad . \quad . \quad . \quad . \quad . \quad (22)^*$$

In which $t$ = thickness in in., $p$ = pressure in lb. per sq. in. (four times the compression approximately), $d$ = diameter in in., $f$ = the allowable tensile strength, which may be considered 16,000 lb. per sq. in., and 0.1 is a constant.

* See Unwin's "Machine Design."

In actual practice, however, the cylinder wall is made somewhat heavier than the above formula would seem to indicate, in order to produce a good casting, to guard against the jacket core falling and making the wall too thin on one side, and to allow for reboring. If the cylinder wall is cast too thin, water from the jacket is liable to pass through into the cylinder. In actual practice the cylinder wall, under the water jacket, varies in thickness from $t = 0.1\ D$ to $t = 0.05$ D, in which $D$ is the diameter of the cylinder bore. The cylinder wall $t_1$, from the water jacket to the flange, may be made $1.2\,t$. Automobile manufacturers use the lightest cylinder permissible and it is common to find a cylinder of 5-in. bore with a 0.25-in. wall. Some foreign concerns manufacture a cylinder of steel casting somewhat lighter than the above dimension, and claim that they produce uniform results in casting. Stationary engines, of course, tend toward the use of heavier cylinders, weight being no especial drawback but on the other hand an advantage up to a certain limit.

The depth of the water jacket $j$ is a dimension over which there seems to be some question, and as a matter of fact, unless the capacity of the circulation pump be considered, it would be impossible to arrive at even an approximate value, and, after having derived a somewhat complex formula, taking into consideration the conductivity of the cylinder walls, the probable heat of combustion, and the temperature of the inlet water, we would find that it would not meet all conditions. The best value for the depth of water jacket would be an average of the practice of different manufacturers. If then a circulation pump with a regulating valve in the supply pipe be used, the amount of entering water may be nicely regulated until just enough water passes through the jacket to make the temperature of the discharge constant at from 160 to 170 deg. fahr.

It is good practice to make the depth of the water jacket

$$j = 0.16\ D \text{ to } 0.10\ D \quad . \quad . \quad . \quad . \quad . \quad (23)$$

We find that the automobile engines use the larger value of $D$, due to the limited amount of circulating water carried and the fact of its being but partially cooled after passing through the jacket. The capacity of the circulation pump is also made small for the same reason, in order that the water in the radiator may have as much time as possible in which to become cooled.

The outer wall of the jacket should be made no thicker than the designer may consider necessary to produce a good casting; in small cylinders this thickness would be relatively greater than in engines of larger bore.

It is probably safe to make the thickness $t_1 = 0.04\ D$ for engines up to 5 or 6-in. cylinder bore, and 0.03 D, or even less, for larger cylinders. In engines using the copper water jacket the thickness is generally about $\frac{1}{32}$ in. Provision must be made in this case for the unequal expansion of copper and iron. This may be accomplished by means of corrugations. Fig. 47, or by the use of an expansion joint. The first-mentioned method is effective and probably the least expensive to use. The copper jacket must be calked into place with lead and oakum, and the head fastened securely in place with proper gaskets. With the copper jacket the cost of the cylinder casting is reduced, but the saving in cost is more than offset by the additional cost of the jacket and the necessary labor required to place it on the cylinder.

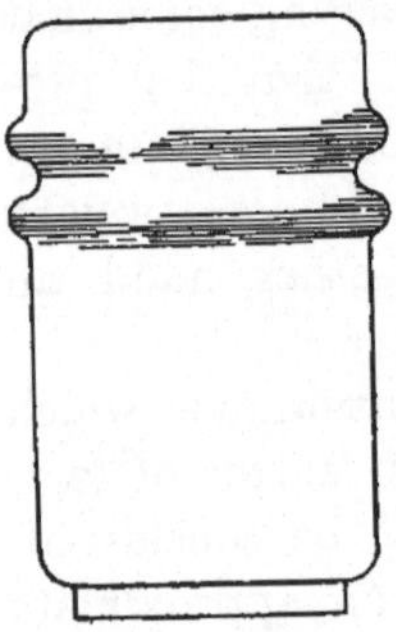

Fig. 47. Copper Water Jacket.

The inlet water should enter the jacket at its lowest point, and the outlet should be placed at the highest point and on the side opposite the inlet if the engine be of the vertical type. If the water leaves the jacket at a point lower than the height of water in the jacket, a pocket is formed in which water stagnates and produces steam — a hot spot in the cylinder is the result. As the pocket so formed is usually on the top of the cylinder, which point is heated the most, it is liable to produce premature explosions and rapidly to deteriorate the cylinder itself.

The water jacket should completely cover that part of the cylinder subjected to the heat of the burning gases or it should continue from the top of the cylinder down to the position which the top of the piston would assume when the engine was on forward dead center. Some designers are inclined to make it even shorter than this, contending that the heat, when the piston has reached the end of its outward stroke, is not sufficient to require a jacket. As previously mentioned, it is good practice completely to jacket the valve box if the valves are placed in pockets, and if the valves are in the head their complete jacketing is a natural consequence.

The cylinder should be relieved as shown at *b*, Fig. 46, in order to facilitate the boring, allowing the tool to run out, and to guard against the piston rings wearing a shoulder in the cylinder: a condition which will result in a knock. To accomplish this latter condition, the cylinder should be relieved at such a point that the top piston ring, when the engine is on inner dead center, will pass half by the edge *b*. Likewise, if the piston carries an oil ring, which is the case in all well-designed engines, this ring should pass half by the edge *c* when the engine is on the outer dead center.

Failure to observe these conditions will result in the lower ring dropping off into the crank case, if it entirely passes the edge *c*, or the upper ring, if it entirely passes edge *b*, will be pulled off in the compression chamber; the result of either of these conditions may be readily imagined. On the other hand, if the travel of the piston does not allow either the upper or the lower ring to partly overtravel edges *b* and *c* respectively, the end of their travel will, in time, be marked by a shoulder and, as the parts of the engine become worn, "knocking" will result.

$$D_1 = D + \tfrac{1}{8}\text{ in.} \qquad (24)$$

The size openings for the inlet and discharge water naturally varies in different makes of engines and under different conditions. If a tank and cooling tower be used,— and this is by far the most desirable method, as the cooling water becomes "broken" and has no lime or sediment to deposit in the jacket,—it is necessary that the openings be made somewhat larger than if the engine were cooled with water at hydrant pressure. For the ordinary stationary engine a water inlet equal in area to a ⅞-in. wrought-iron pipe is sufficient, while the outlet should equal in area a 1-in. pipe. The outlet pipe should, in all cases, be the larger in order to allow for the expansion of the water due to the heat absorbed. Automobile and marine engines use somewhat smaller openings, it being good practice to make the inlet ¾ in. and the outlet ⅞ in.

In case there is a tendency for the water to pass directly through the jacket without cooling all parts of the cylinder

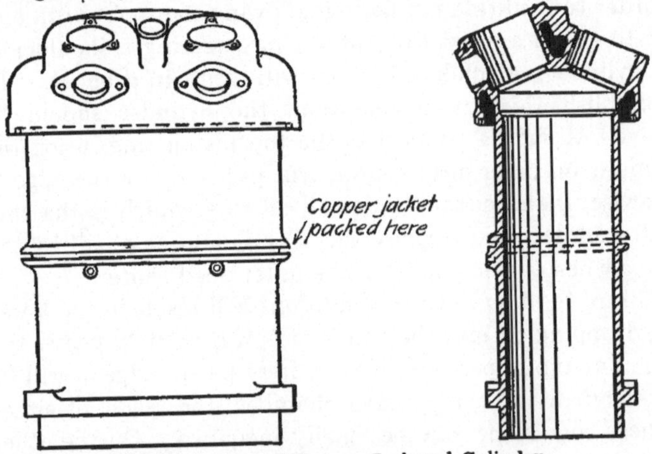


**Fig. 48. Type of Copper Jacketed Cylinder.**

wall, baffle plates, for deflecting the water into its proper channel, should be cast in.

The boring of the cylinder is an operation in which extreme accuracy is necessary, and the machines must be heavy enough to prevent all chattering or vibration. The best engines have their cylinders bored and ground, as when this is done it takes considerably less time to "work in the cylinder." When an engine is marketed with the cylinder finished by boring alone, it is always found to be true that, after using the engine for a number of months, its power increases as the cylinder walls and piston rings are worn to a glassy smooth surface, a condition productive of least friction and the highest compression. Bore the cylinder to an even decimal and make the piston smaller, as will be described later.

The cylinder illustrated in Fig. 46 is designed to be fastened to the frame or crank case by means of four corner bolts, or cap screws, passing through holes *d*. The size of these bolts may be found from the formula

$$d_b = D \frac{p}{fn} \quad \ldots \ldots \quad (25)$$

In which

$d_b$ = diameter of bolt at root of threads.
$D$ = diameter of cylinder.
$p$ = maximum pressure.
$f$ = allowable unit stress. (About 10,000 lb. per sq. in.)
$n$ = number of bolts.

This same formula should be used in determining the size of studs or bolts to be used in the cylinder head if it is cast separate from the cylinder proper, as shown in Fig. 40. For the cylinder head, however, $f$ should not be taken at more than 8,000 lb. per sq. in., due to the fact that initial stress is probably introduced in tightening the bolts hot, as must be done to insure a tight joint.

If the jump-spark ignition system be used, the spark-plug hole *e* should be tapped to receive the plug; 0.5-in. pipe tap is standard for most spark plugs.

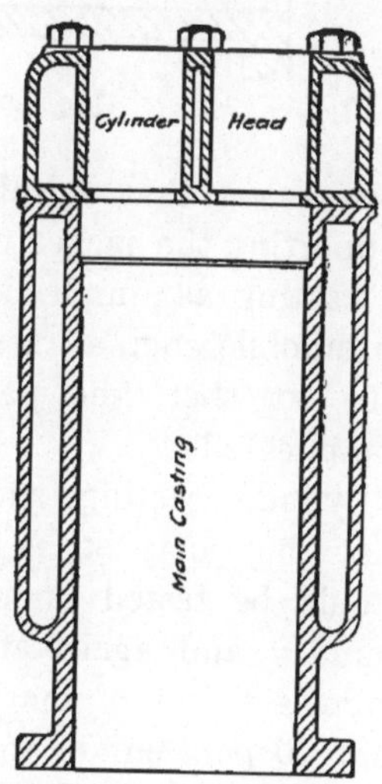

Fig. 49. Water-cooled Cylinder with Separate Head.

In the cylinders shown in Figs. 48, 49, and 50 the same general dimensions may be adhered to and for that reason they are not discussed. They are inserted in this work in order to give the designer some idea of the different forms of cylinders, and while not, by any means, fully covering all conditions, they should offer sufficient suggestions to enable the average draftsman to design a good working cylinder.

Cylinders with the water jacket and the head cast integral form, at best, an expensive casting, usually ranging in cost from eight to fourteen cents a pound. It is imperative for

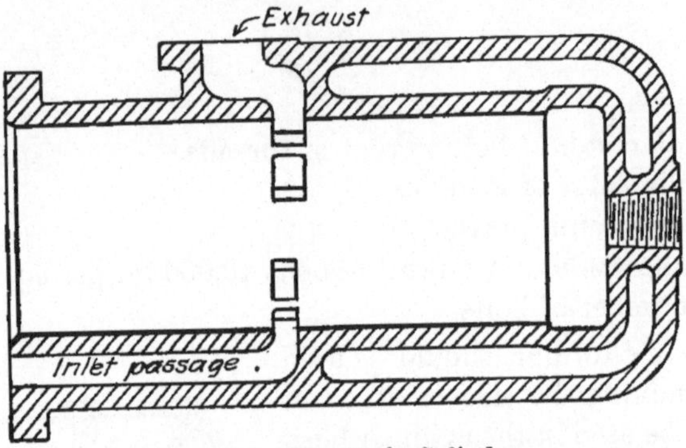


Fig. 50. Two-Cycle Cylinder.

this reason that suitable means be provided for properly supporting the jacket core and thereby making the problem of casting as simple as possible. Cylinders, especially for automobile engines, are often cast in pairs or even three or four together, the patterns, in consequence, being very complicated.

Cylinder castings should be of the best gray iron in order that they may be as tough as possible, and each casting should be tested under water pressure before leaving the foundry and again after machining. One of the leading manufacturers of marine engines has made the statement that 20 per cent of the finished cylinders in their shop are found to be so faulty that they are "scrapped."

# CHAPTER XVI.

## THE FLYWHEEL.

THE function of the flywheel is to store up energy imparted to it during the impulse portion of any cycle, and to restore this energy thus received to the machine during the remainder of its cycle of operations, or until it receives another impulse.

It is possible to so design a flywheel that it will reduce to a minimum amount the fluctuation of speed of the engine.

Rankine treats the fluctuation of speed as a coefficient of fluctuation and designates this coefficient, $\frac{\Delta E}{2\,E_0}$. In which $\Delta E$ is the excess of energy received during impulse, and $E$ is the actual energy. A flywheel reduces this coefficient of fluctuation to a fixed amount, which for ordinary machinery is generally taken at $\frac{1}{32}$, and for finely regulated machinery $\frac{1}{50}$ or $\frac{1}{60}$; that is to say, the variation in speed may be from about 1.5 per cent to 3 per cent, depending on the nature of the machine. An engine, then, whose speed on an average is 200 rev. per min., with a variation of 3 per cent, will vary between highest and lowest speed $200 \times 0.03 = 6$ rev. per min., and, since 200 rev. per min. is its average speed, the speed would vary between 197 rev. per min. and 203 rev. per min.

The speed regulation as effected by the flywheel must not be confounded with the governing of the engine. Illustrating this: suppose the speed of the engine at no load was 205 rev. per min., while at full load its speed was 195 rev. per min. Its speed variation between these two loads might then be considered as $\frac{205 - 195}{200} = 5$ per cent. But the regulation of this variation, plainly, could not, in any way, be effected by the weight of the flywheel but should be controlled by the governing apparatus (Chapter XXII).

The steadiness of speed obtainable in an engine is not dependent on the flywheel alone. In a direct-connected electric-generating unit the armature is a potent factor in speed regulation, as are also the pulleys, shafting, and other machinery, when belt connected. Some direct-connected units utilize an auxiliary flywheel on the tail shaft of the dynamo. As the balancing effect of the flywheel is determined, most largely, from its rim, which is the heaviest portion as well as farthest from its axis of revolution, it is common practice to disregard the weight of the hub and arms, or web, as a balancing factor, although they exert their weight but in lesser extent. For this reason the web or arms should be made as light as possible to resist the strain to which they are subjected. It is obvious that the proportion of idle strokes to the actual strokes is so large, in a gas or gasoline engine, even a two-cycle engine under most favorable conditions getting only one impulse per revolution where a steam engine receives two, that the weight of the flywheel must be comparatively large.

The only method of actually calculating the weight of the flywheel required is to make a graphical diagram showing the resultant forces of the cylinder as acting on the crank pin, together with the effect of the reciprocating parts, and from the force diagram thus derived to determine the weight of the wheel which will equalize the unbalanced portions. This method is tedious and complicated and, while serving to show clearly the theory of the flywheel and acting as a basis for the empirical formulas in common use, it has no very great value to the practical designer. For a detailed description of this method the reader may refer to a small pamphlet on "Dynamics of Reciprocating Engines," as revised from the *Michigan University Technic* of 1888. The empirical formulas used in general practice give results that are accurate to within a few per cent of those which would be obtained from an actual diagram. The *American Machinist* gives the following formula:

$$W = \frac{aAS}{R^2D^2} \qquad . \quad . \quad (26)$$

In which $W$ = weight of rim, $a = 24{,}000{,}000$ for engines firing charge every revolution and 48,000,000 for engines of the four-cycle type,

$A$ = area of piston in sq. in.
$S$ = stroke in ft.
$R$ = rev. per min.
D = outside diameter of wheel in ft.

The revolutions per minute or speed for which engines are designed being known, the above formula, with the constant $a$ as given, is good for average purposes, but, as we have observed, the value of $\frac{\Delta E}{2E_0}$ may vary, for different purposes, from 1.5 per cent to 3 per cent; hence it is obvious that for varying conditions we should introduce a factor to take care of this allowable variation. As a matter of fact the constants $a = 24{,}000{,}000$ and $a = 48{,}000{,}000$ are average figures, and the value of $a$ should be taken from 20,000,000 to 30,000,000 and from 40,000,000 to 60,000,000. We are then able to modify the formula to suit different conditions as follows:

$$W = \frac{aAS}{R^2D^2} \cdot x \quad . \quad . \quad . \quad . \quad . \qquad (27)$$

In which we consider $a$ as 20,000,000 in all cases for a two-cycle engine, and as 40,000,000 in all cases for a four-cycle engine, and $x$ is a constant, for different conditions, as shown in the following table:

| | |
|---|---|
| Portable engines | 1.0 |
| Pumping and ordinary use | 1.1 |
| Driving machine tools | 1.2 |
| Driving looms or textile machinery, etc. | 1.3 |
| Driving electric machinery, etc. | 1.4 |
| Driving cotton spinning, etc. | 1.5 |

Automobile and motor-boat engines permit, and their usage demands, a lighter flywheel than for stationary purposes, and in these engines we may safely use a value of $x$ as low as 0.60 for automobiles and 0.75 for motor boats.

Having determined the weight of rim necessary, its content in cubic inches is determined by dividing the weight by

0.27, which is the approximate weight, per cubic inch, for cast iron. Now assume some thickness for the rim — the thickness should, wherever possible, be its minor dimension, so that its center of gravity may be as far from its center of rotation as possible. The mean diameter of the rim will, then, be equal to its outside diameter plus its inside diameter, divided by two. The mean diameter and circumference and thickness of rim being known, its width must be such that it may contain the number of cubic inches, above determined, necessary to produce the required weight. Use the nearest $\frac{1}{8}$ in. above the calculated width.

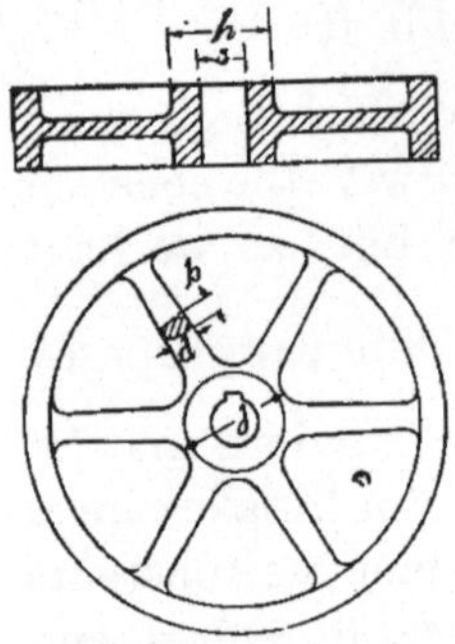

Fig. 51. Flywheel.

Most flywheels used on stationary engines are made with six spokes, while the majority of automobile and marine engines use the webbed pattern. Fig. 51 is a spoked wheel showing the average dimensions in good use. In the equations, $s$ = crank-shaft diameter.

Then

$$h = 2\,s.$$
$$j = 3\,s.$$
$$d = s \text{ (approximately).}$$
$$b = 0.4\,s \text{ (approximately).}$$

The dimension $d$ may be computed from Unwin's flywheel formula, in which

D = diameter of pulley in in.

$B$ = breadth of rim in in.

$n$ = number of arms,* thus:

$$d = 0.6337\sqrt[3]{\frac{BD}{n}} \text{ (Single belt)} \quad . \quad . \quad . \quad . \quad (28)$$

$$d = 0.798\sqrt[3]{\frac{BD}{n}} \text{ (Double belt)} \quad . \quad . \quad . \quad . \quad (29)$$

* For a flywheel transmitting no power a smaller value may be used.

Some manufacturers design their flywheels in two parts, and provide notches, in the rims, in which a bar may be inserted for turning the engine over. Fig. 52 represents such a wheel; the outer rings *a* and *b* are shrunk on, leaving the space in which the notches have been cut, as shown. The rings and bolts bind the wheel firmly together.

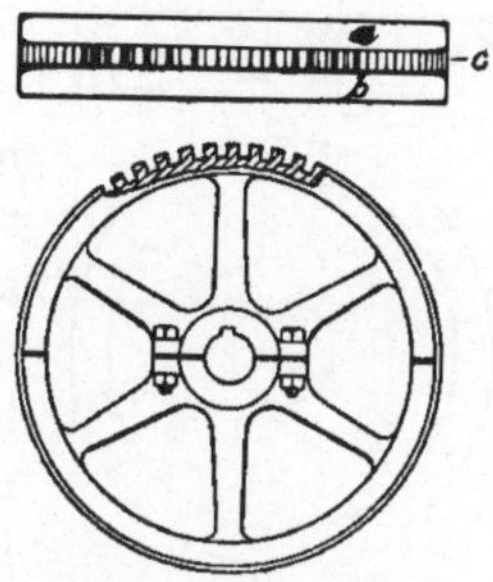

Fig. 52. Flywheel.

The maximum velocity of the rim is generally taken at 6000 ft. per min. Then $6000 = \pi RD$.

Whence

$$D = \frac{6000}{\pi R} = \frac{1910}{R} * \qquad (30)$$

The size of key necessary may be found in Tables X to XIII and should be of as strong cross-section as possible and a tight drive fit in order to lessen the tendency to work loose.

TABLE X.

DIMENSIONS OF FLAT KEYS, IN INCHES.

| Diam. of shaft.... | 1 | $1\frac{1}{4}$ | $1\frac{1}{2}$ | $1\frac{3}{4}$ | 2 | $2\frac{1}{2}$ | 3 | $3\frac{1}{2}$ | 4 | 5 | 6 | 7 | 8 |
|---|---|---|---|---|---|---|---|---|---|---|---|---|---|
| Breadth of keys .. | $\frac{1}{4}$ | $\frac{5}{16}$ | $\frac{3}{8}$ | $\frac{7}{16}$ | $\frac{1}{2}$ | $\frac{5}{8}$ | $\frac{3}{4}$ | $\frac{7}{8}$ | 1 | $1\frac{1}{8}$ | $1\frac{3}{8}$ | $1\frac{1}{2}$ | $1\frac{3}{4}$ |
| Depth of keys.... | $\frac{5}{32}$ | $\frac{3}{16}$ | $\frac{1}{4}$ | $\frac{9}{32}$ | $\frac{5}{16}$ | $\frac{3}{8}$ | $\frac{7}{16}$ | $\frac{1}{2}$ | $\frac{5}{8}$ | $\frac{11}{16}$ | $\frac{13}{16}$ | $\frac{7}{8}$ | 1 |

TABLE XI.

DIMENSIONS OF SQUARE KEYS, IN INCHES.

| Diam. of shaft......... | 1 | $1\frac{1}{4}$ | $1\frac{1}{2}$ | $1\frac{3}{4}$ | 2 | $2\frac{1}{2}$ | 3 | $3\frac{1}{2}$ | 4 |
|---|---|---|---|---|---|---|---|---|---|
| Breadth of keys....... | $\frac{5}{32}$ | $\frac{7}{32}$ | $\frac{9}{32}$ | $\frac{11}{32}$ | $\frac{13}{32}$ | $\frac{15}{32}$ | $\frac{17}{32}$ | $\frac{9}{16}$ | $\frac{11}{16}$ |
| Depth of keys......... | $\frac{3}{16}$ | $\frac{1}{4}$ | $\frac{5}{16}$ | $\frac{3}{8}$ | $\frac{7}{16}$ | $\frac{1}{2}$ | $\frac{9}{16}$ | $\frac{5}{8}$ | $\frac{3}{4}$ |

TABLE XII.

DIMENSION OF SLIDING FEATHER-KEYS, IN INCHES.

| Diam. of shaft..... | $1\frac{1}{4}$ | $1\frac{1}{2}$ | $1\frac{3}{4}$ | 2 | $2\frac{1}{4}$ | $2\frac{1}{2}$ | 3 | $3\frac{1}{2}$ | 4 | $4\frac{1}{2}$ |
|---|---|---|---|---|---|---|---|---|---|---|
| Breadth of keys ... | $\frac{1}{4}$ | $\frac{1}{4}$ | $\frac{5}{16}$ | $\frac{5}{16}$ | $\frac{3}{8}$ | $\frac{3}{8}$ | $\frac{1}{2}$ | $\frac{9}{16}$ | $\frac{9}{16}$ | $\frac{5}{8}$ |
| Depth of keys..... | $\frac{3}{8}$ | $\frac{3}{8}$ | $\frac{7}{16}$ | $\frac{7}{16}$ | $\frac{1}{2}$ | $\frac{1}{2}$ | $\frac{5}{8}$ | $\frac{3}{4}$ | $\frac{3}{4}$ | $\frac{7}{8}$ |

* Kent's "Mechanical Engineer's Pocket Book," pages 821-822.

Table X should be used for all fixed work wherein the key not only drives, but also holds the parts against endwise motion. These keys are tapered and "bear all over." *

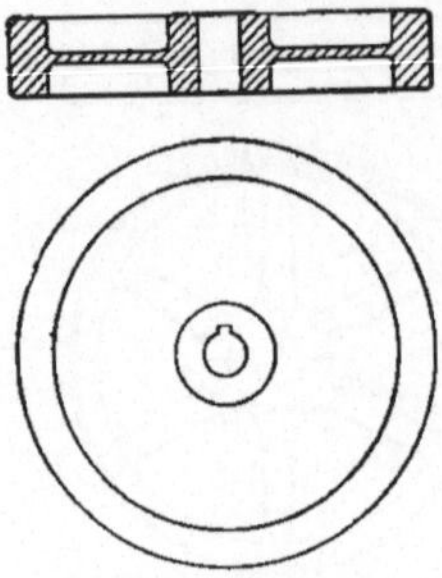

Fig. 53. Flywheel.

A webbed flywheel, Fig. 53, should use a comparatively small thickness for its web. This thickness varies, in automobile and marine engines, from 0.5 to 0.75 in., and need not be computed, as, in small wheels, a web as thin as could safely be poured would probably be as strong as six spokes. In most cases the thickness of the web will be found to be approximately one-fourth the shaft diameter. In a spoked wheel, always fillet the spokes well where they join the rim, as the large mass of metal may, otherwise, cause a crack to form at the joint.

The flywheel must be accurately turned and balanced. If, after finishing, one side is found to be heavier than the other, holes are drilled in the heavy side in order to overcome this difficulty. Some manufacturers go so far as to finish the face of the web in a webbed wheel, but except for finest machines this will be found unnecessary.

* John Richards, in *Cassier's Magazine*.

# CHAPTER XVII.

## THE FRAME.

THE gas-engine frame serves a twofold purpose; it rigidly supports the crank shaft and cylinder and resists their tendency to separate from the force of the explosion, and, by its inertia, absorbs a part of the unbalanced forces in the engine and transmits to the foundation proper the remainder of these forces.

The advantage of having a comparatively heavy foundation on which to rest the engine frame is due to the fact that it is undesirable as well as impractical to make the frame heavy enough in itself to effectively absorb the vibrations, as such a large mass of metal would be exceedingly cumbersome to handle. In view of this fact it is customary and good practice to make the engine frame of cast iron and as heavy as may be, at the same time preserving its mechanical appearance and limiting its weight to such a degree as to permit of its being handled in the machining process. For these reasons it is not customary for the engine designer to compute any part of the frame, with the exception of the bolts, as it is assumed that it will be amply strong to resist any shock to which it may be subjected.

The formula to be used for computing the bolts that fasten the cylinder to the frame is given in Chapter XV, page 135, and needs no further discussion.

*Horizontal Engines.* — Fig. 54 will serve to show the general character of design of frame as used in gas engines. Nearly every manufacturer will be found to modify his frame to some extent, and it would be useless to go more fully into their special characteristics.

In the frame shown in Fig. 54, the bearings are set at a 45° angle, in order that most of the thrust of the crank shaft may be against the frame instead of against the bearing studs. While an angle of 45° will not be such that the maximum forward thrust will be received by the frame, a greater angle would not be desirable, as it would bring the joint in the brasses too near the bottom of the bearing. In

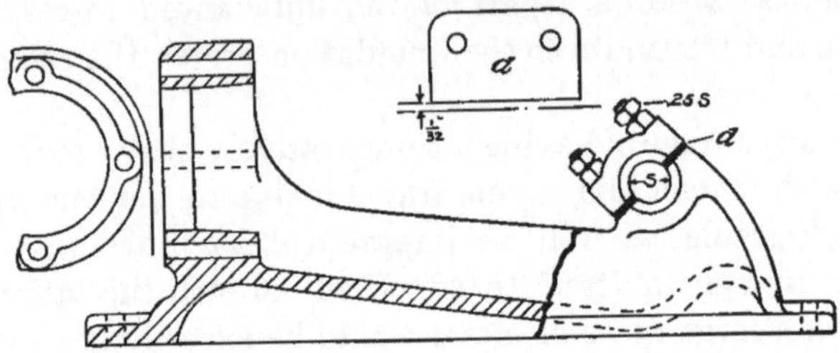


Fig. 54. Frame for Horizontal Engine.

the design shown brass liners, *a*, are used to take up the play in the shaft. Four or five of these liners, made of light-gage brass, should be used. They are cut to conform to the shape of the cap and to clear the shaft $\frac{1}{32}$ in. Two or three may be made of 30-gage B. & S. and a like number of about 40-gage.

A good rule to follow for the size of the bearing studs is to make their diameter, at the base of thread, 0.25 the diameter of the shaft.

Some manufacturers carry a projection of the frame out under the cylinder, on which lugs are cast, and bolts are passed through corresponding lugs on the cylinder. It is doubtful if this is good design, as the unequal expansion of the cylinder and frame will either loosen these bolts or cause a distortion of the cylinder alignment.

The air supply for this type of engine is generally drawn from the hollow base, in order to make the operation as quiet as possible and to secure a warm dry suction.

*The Vertical Engine.*—In this type of engine, the frame usually becomes a crank case, completely housing the shaft and cam gearing. There are exceptions, of course, in cheaply constructed single-cylinder engines for general purposes, but the modern high-class vertical engine as a rule has a closed crank chamber. The same general rules apply to proportioning and designing as in the horizontal type, whether the frame consists of a closed case or is of similar construction to that used in the horizontal engine.

In the crank-case type of engine the case is generally made in two halves, the lower half carrying the bearings, which are independent of the upper half. The cylinders are

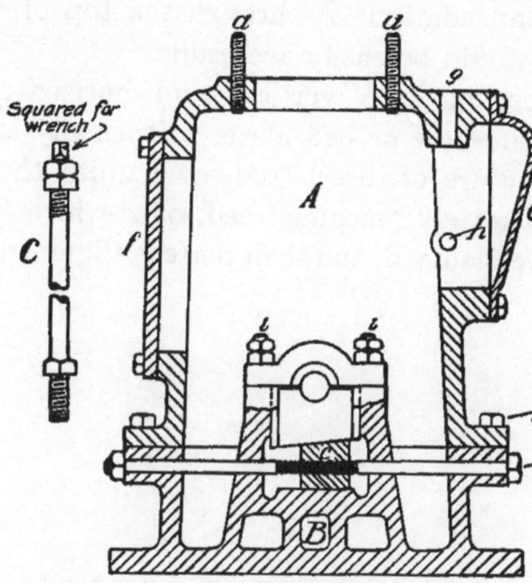


Fig. 55. Frame for Vertical Engine.

fastened to the upper half, and the two halves are fastened securely together.

Fig. 55 gives a good general idea of this class of design. The formula for the cylinder studs, *a*, has already been given and needs no further discussion. The two halves of the case are shown fastened together by means of the bolts or cap screws, *b*. The bearing pedestal, as shown, is integral with the lower half. The play in the bearings is taken up by means of the wedge, *c*, which acts on the screw, *d*, as shown. As the bearings in this case are more or less inaccessible, it is desirable that this method of taking up the play be used instead of placing liners between the bearings, as described in the horizontal type. Some of the best horizontal engines also use the wedge and screw to secure their adjustments.

Access is had to the interior of the case by means of plates *e* and *f*. Plate *f*, as shown, covers the cam and valve, as well as the starting mechanism, if used; the push-rod bushing being inserted at *g*, above the cam shaft, shown at *h*. The bearing studs *i*, as shown, require that the side plate *f* be removed in order to reach them. A better design would be to use a long cap screw *C*, with a shoulder to act on the bearing cap, and continue it through the top of the crank case, where it would be easily accessible.

Manufacturers of both vertical and horizontal engines often make a sub-base or bed plate. Especially is this true in the manufacture of direct-connected units, the sub-base forming an accurately machined bed, onto which the engine and dynamo are fastened, and their perfect alignment assured.

# CHAPTER XVIII.

## ENGINE FOUNDATIONS.

THE engine foundations are almost always built by the owner or contractor from drawings furnished by the engine builder. The drawings consist of accurately dimensioned plates showing location of the foundation bolts, and also a drawing of a template to be made for locating these bolts in the foundation.

Without a good foundation an engine is bound to give trouble, sooner or later, from settling or from the engine working loose from the foundation and getting out of line.

The foundation of the engine may be considered as a part of the bed plate or frame, in that it should be given sufficient mass to absorb, by its inertia, the effect of the suddenly applied cylinder or crank-pin forces not absorbed by the bed plate and frame. It should get a good bearing on solid ground, the quality of the soil governing, to a large extent, the depth of foundation necessary. Under no conditions should an engine be fastened directly to the floor of the engine room, except for a temporary job, and furthermore, the floor of the building should be absolutely independent of the foundation, in order to prevent the transmission of vibrations to other parts of the building, except as may be transmitted by the concussion of the air or the vibration of the ground. When it becomes necessary to reduce the vibration still further, a layer of some insulating material, 2 or 3 in. thick, should be placed between the foundation and the surrounding soil. A substance such as deadening felt, horsehair felt, or cork may be used and a layer 8 or 10 in. thick placed next to the foundation. Cork, however, should not be used below the foundation owing to its tendency to absorb water and swell. A bed of dry sand as a bottom has been used with good results.

If it becomes necessary to install an engine on an upper floor where a bearing on the ground cannot be obtained, a crib of heavy timbers should be built below the floor to support a foundation of concrete. This foundation will absorb, by its inertia, a great deal of the vibration.

An engine foundation may be made of concrete, brick laid in cement, or stone. If of concrete, the mixture should be one part good Portland cement, two parts sharp clean sand, and four parts broken stone small enough to pass through a 2-in. ring. The concrete should be laid in layers of not more than 6 in., each layer being thoroughly tamped before the next layer is put in. Foundations made of brick should be laid in cement mortar composed of one part Portland cement to two parts clean sharp sand. The brick used should be hard-burned foundation brick. A very good foundation, up to the floor line, may be made of brickbats laid in the above-mentioned mortar, care being taken to fill all voids. This foundation may then be built up of concrete or brick, in the ordinary way, to receive the engine. A brick foundation should have a cap of limestone or granite, or a cement cap 1 ft. in thickness may be put on. Stone foundations should be laid up in cement mortar, and in regular steps, being sure that the stones have a good level bearing, so as to prevent any tendency to slide. A concrete foundation should be given a batter of about 2 or 3 in. to the foot, and a brick or stone foundation should follow, in general, the same outline. A concrete foundation of standard design is shown in Fig. 56.

If, as frequently is the case, the engine is a direct-connected unit, that is, an engine directly connected to a dynamo, then the foundation should be so designed as to take in the dynamo and outboard bearings.

The weight of an engine foundation is seldom computed, as differently designed engines are subject to different degrees of shock, from their unbalanced forces, and the weight of their frame and bed plate may vary to such an extent that any formula would be applicable to but few cases were the surrounding conditions the same. With the condition of the soil varying, it would be impossible to derive a universal formula that could be used in all cases with even a reasonable degree of accuracy. When the engine and building foundation both rest on rock, it is often found that this rock carries the vibration of the engine to the building. It is good practice to make the foundation

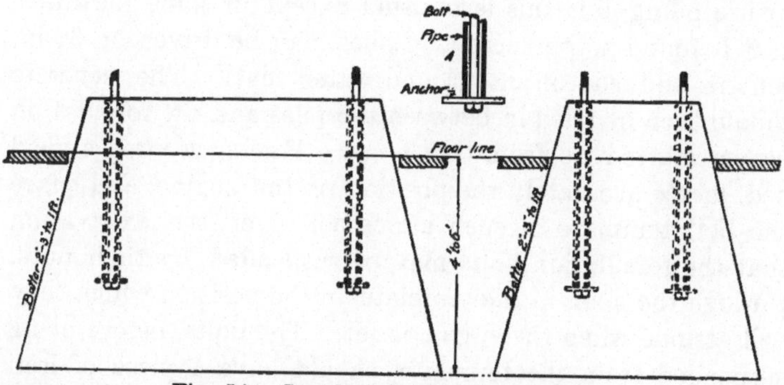

Fig. 56. Standard Engine Foundation.

amply heavy; go well below the frost line and give the foundation a good broad base and its weight will in most cases be ample. If brick is used in the foundation, it is well to remember that 1.25 cu. ft. of brick is about equal to 1 cu. ft. of concrete. (See table of weights.)

The number of foundation bolts used varies with the size of the engine, in no case being less than four and usually not more than eight. The approximate number may be found by the formula

$$N = \frac{P}{6}, \quad (31)$$

Where $P$ is the horsepower of the engine always use the nearest even number of bolts.

The size of the bolts used varies from 0.75 in. for small engines to 1.5 in. or 1.75 in. for large machines. It is not advisable to make the bolts smaller than 0.75 in., as there is a good chance of their being twisted off in tightening. The engine frame should be, to all intents and purposes, a part of the foundation, so that their combined inertia may act together to absorb the shocks. For this reason the foundation bolts should be ample in size and of sufficient number to accomplish this result.

In laying out a foundation proceed as follows: Dig down to good firm ground, or hard pan, and level this bottom off carefully; in case the ground is marshy it may be necessary to use piling, but this is unusual except in some localities. If it is found to be necessary, piles may be driven on 30-in. centers and cut off at ground-water level. The concrete should then be filled in between the piles and the foundation proper started up from that point. Having secured a good bed, locate accurately the position of the engine, and place the bolt template in such a position over the excavation that the foundation bolts may be suspended, by their nuts, through the holes in the template, in the position which they will assume when the engine is set. The bolts, before being suspended from the template, should have a piece of iron pipe slipped over them, Fig. 56, *A*. This pipe should be large enough to allow a slight movement of the bolt in setting the engine. The pipe is supported at its lower end by the bolt anchor, and its upper end should come just below the top of the finished foundation. The foundation may now be built up around these pipes. When the top course, either brick or concrete, is laid, be sure that the surface which is to receive the bed plate is perfectly level. This leveling should be done with a Y level and must be accurate. After the engine has been set the pipes should be filled with cement mortar. A good size of pipe to use is a 2-in. standard gas pipe.

Foundations after being built should be allowed to set for at least a month, unless circumstances demand them to be used sooner, in which case the concrete should be made as dry as possible to secure a good mixture.

It should not be assumed that it is common practice to insulate the foundation with a shock absorbent, as above described. This is done only when the location of the engine is such as to make it desirable that every precaution be used to prevent a nuisance.

# CHAPTER XIX.

## THE CRANK SHAFT AND RECIPROCATING PARTS.

THE crank shaft, connecting rod, and piston, together with their respective parts, constitute the rotative and reciprocating parts of a gas engine. The crank shaft, as previously described, carries the flywheel, and balance weights, if used.

With but few exceptions, gas-engine pistons are of the trunk variety. The reason for this very general use of the trunk piston is that if the engines were provided with a piston rod and made double-acting, as is the practice in steam engines, the rod end of the piston would be subjected to as much heat as the head end, and the rod, becoming heated, would expand and cut into the stuffing-box. Except in the case of two-cycle engines, in which the crank case is used as a primary compression chamber, there is no necessity of closing the forward end of the cylinder, the piston itself being made long enough to act as a guide without the use of a piston rod and cross-head. A few engines have been designed which were double-acting, the rod and packing box being cooled by means of water made to circulate about them, but as these types may be considered in the "freak" class no special discussion will be given them.

The crank shaft of a gas engine, while it serves in the same capacity as the shaft of a steam engine, must be made very much stronger, since it is subjected not only to the strain due to the conversion and transmission of the power received but, at the beginning of the working stroke, it is subjected to six times the average power of the engine.

In determining the size of crank shaft necessary, the diameter of the piston and the maximum cylinder pressure are taken as the basic conditions for the computation.

Let $D_s$ = the diameter of the shaft in in.
$D_c$ = the diameter of the cylinder in in.
$P_m$ = the maximum cylinder pressure.

For a drop-forged steel shaft the average practice is to make

$$D_s = 0.05\, D_c \sqrt[3]{P_m} \quad . \; . \; . \; . \quad (32a)$$

For malleable or wrought iron,

$$D_s = 0.06\, D_c \sqrt[3]{P_m} \quad . \; . \; . \; . \; . \; . \quad (32b)$$

These values will be found to produce results closely in keeping with average practice when the stroke is equal to 1.5 $D_c$. E. W. Roberts gives the following formulas for crank shafts where the ratios of the stroke $L$ to the cylinder bore is other than 1.5 : 1.

For wrought iron,

$$D_s = 0.056 \sqrt[3]{P_m L D_c^2} \quad . \; . \; . \; . \; . \quad (33a)$$

For steel,

$$D_s = 0.052 \sqrt[3]{P_m L D_c^2} \quad . \; . \; . \; . \; . \quad (33b)$$

As he remarks, the formulas will be found to give results larger than the average practice, but for stationary engines this is a good fault. For marine and automobile engines the values will be found to be much larger than average practice requires.

The length of the crank-shaft bearing necessary is readily determined, once the diameter is known, from the following formula:

$$l = \frac{D_c^2\, P}{1018\, D_s} \quad . \; . \; . \; . \; . \; . \; . \quad (34)$$

In which $l$ = length of the bearing.
$D_c$ = cylinder diameter, as above.
$D_s$ = Shaft diameter, as determined.
$P$ = Mean effective pressure.

In practice, however, the crank-shaft bearings are usually made much longer than the value which would be obtained by the use of this formula. A good rule to follow is to make $l = 2.5\, D_s$.

The crank-pin diameter is usually made from 1.1 to 1.3 the diameter of the crank shaft, and its length would then be determined from the following formula:

$$l_p = \frac{D_c^2 P}{508 \, D_p} \quad . \quad . \quad . \quad . \quad . \quad . \quad . \quad (35)$$

In which $D_p$ = diameter of the crank pin.

$l$ = length of the crank pin.

$D_c$ and $P$ as in the above formula.

The crank arms should be sufficiently large to withstand the suddenly applied forces to which they are subjected. Fig. 57 illustrates a good design of crank for a stationary slow-speed engine. The relative dimensions of the crank

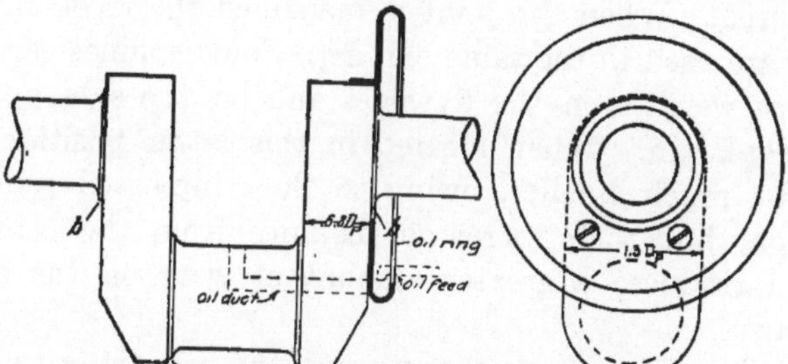


Fig. 57. Crank with Oil Ring.

arms are given in terms of the crank-shaft diameter, and will be found to produce values closely in keeping with stationary-engine practice, although slightly heavy for automobile or marine engines. It is customary to leave $\frac{1}{32}$ in. finish for the ends of the crank-pin brasses as indicated at *a*. In larger shafts this finish is sometimes increased to $\frac{1}{16}$ in. A like finish is allowed at *b*. This allows the fillets for the brasses to be finished, as shown, without facing both surfaces of the crank throws.

Single-throw cranks for high-speed engines are provided with balance weights in order to neutralize the weight of the crank arms and that part of the connecting rod regarded as rotative. It is customary to assume two-thirds of the weight of the connecting rod to be centered in the crank pin and hence to be added to the weight of the pin when computing the balance weights necessary. The usual location for the balance weights is to place them one on either crank throw and on the side of the shaft opposite to the pin. When located in this manner they should be securely fastened in place with pins, and a perfect bearing secured by babbitting the joint or by machining both the weights and crank arms to a perfect and tight fit. The slightest play in these weights will increase very rapidly as the engine is run and will cause no end of trouble. When the weights are babbitted a groove in the weight and crank arm, as indicated in Fig. 58, retains the babbitt and secures the weight firmly, and when the rivet and cap screw are in place, as shown, a good job is secured. When the joint is machined the rivet and cap screw are used in the same manner. Some engines carry the balance weights on the flywheel and on the side opposite the crank pin. When located in this latter position they may be much smaller, owing to their increased radius of rotation, but their increased distance from the center of force introduces a greater amount of wear on the engine bearings.

In balancing a single-throw crank the first thing to determine is the weight, considered centered in the crank pin, to be balanced by the counter weight. As previously mentioned, two-thirds of the connecting rod is considered as rotative. Then the weight to be balanced would be equal to $\frac{2}{3} W_r + W_p + x$.

In which

$W_r$ = Weight of connecting rod, brasses, and bolts, etc.

$W_p$ = Weight of crank pin and that part of the arms concentric to the pin.

$x$ = Weight of the remainder of the crank arms considered centered in the crank pin.

The values $W_r$ and $W_p$ are easily determined, but the value of $x$, accurately computed, would possibly require a rather complicated equation. Hence it is customary to determine the moment of $W_r$ and $W_p$ and then to balance them with a weight 10 per cent in excess of the calculated value.

Then $1.1 (W_r + W_p) l$ = moment of balance weights necessary ($l$ being the distance from center of shaft to center of pin).

Now taking the distance of the center of gravity from the crank-shaft center to be $l_1$, the weight necessary to balance the crank and rod would be

$$W = \frac{1.1 (W_r + W_p) l}{l_1} \quad . \quad . \quad . \quad . \quad . \quad (36)$$

Half of this amount in each weight with their centers of gravity $l_1$ distant from the shaft would nearly produce a balance. To determine the value of $l_1$ it is customary to "cut and try." Lay out the crank and balance weights, as shown in Fig. 58. Then cut out a template of cardboard

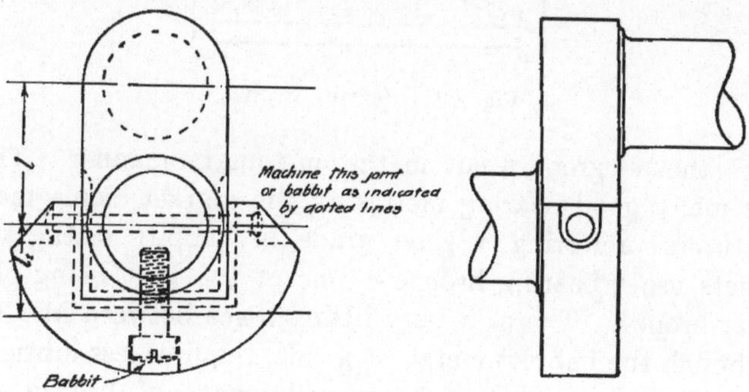


Fig. 58. Balanced Crank.

the exact shape of the weight as drawn; find the center of gravity of this template by balancing it on a pencil point and then determine the distance of this point from the center of the shaft when the weight is in place. Using this value in equation (36) gives the total amount of balance necessary, and half of this amount is to be placed in each weight. If the value of $l_1$, as determined, is found to be so small that, in order to make the two weights equal to $W$, it is necessary to make them thicker than the crank arms, then a larger template should be laid out and a larger value for $l_1$ determined, from which a smaller value for $W$ will be found. Crank shafts with more than one throw are not balanced, since the different rods and arms balance each other; neither is it customary to balance the crank of the cheaper industrial engines.

The more expensive engines have their shafts turned and then ground in order to insure a smooth bearing.

The shaft and connecting-rod bushings are made of some good bearing metal, turned and finished to the proper size, after which they should be "scraped in" and the oil grooves cut to carry the lubricant. Fig. 59 illustrates such a bushing

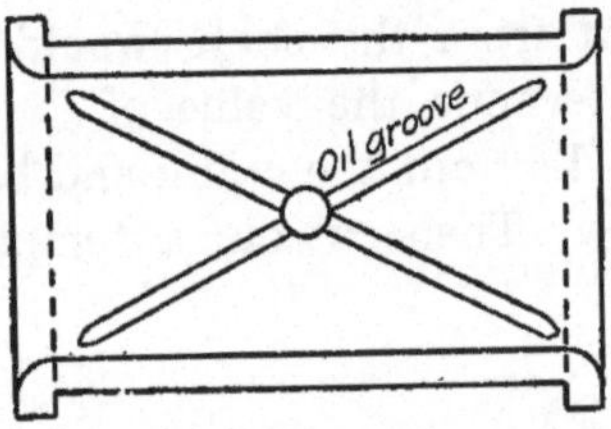


Fig. 59. Bearing Bronze.

with the oil grooves cut in the customary manner. There are many good bearing metals on the market; some manufacturers preferring a good grade of babbitt metal, while others use a bearing bronze — one of the best being phosphor-bronze. There is very little choice between the two, although the babbitt metal, as a rule, requires less lubricant.

One of the very best bearings is made of bronze with babbitt poured in, in order to facilitate the lubrication.

## Oil Rings.

In order to secure good lubrication of the crank pin in closed crank-case engines when splash lubrication is used, means must be provided for conducting the oil contained in the crank case to the pin bushing. This is often accomplished by milling a slot in the end of the rod cap and through the bushing; the rod end passing through the oil during its revolution, enough is picked up and works into the pin through the slot to keep the bearing fairly well lubricated.

A better method, however, and one which is coming into more general use, is to provide an oil ring located on the outside of the crank arm as shown in Fig. 57 at *A*. An oil duct, shown in dotted lines, leads from the ring to the pin. The ring is continually immersed in the oil up to a certain point and the centrifugal force picks up the oil and carries it out into the rim of the ring from which it is forced through the duct to the pin. The oil ring is, as a rule, made of aluminum, securely screwed to the face of the crank arm, and it should be made just heavy enough to produce a perfect casting. Open-crank engines are provided with an oil cup on the connecting-rod end, and no oil ring is required.

### The Connecting Rod.

The connecting rod of a gas engine transmits directly to the crank shaft the energy imparted to the piston. As a general proposition it acts as a strut, the component at right angles to its axis being comparatively small but greatest in long-stroke engines using a short rod. This transverse strain is at its maximum when the crank is midway in the working stroke. Connecting rods are made in several different shapes, the most popular design for stationary engines being of rectangular cross-section, with its greatest dimension at right angles to the crank pin and tapering from the crank end toward the "wrist pin" bearing. While this design does not produce a rod with its stiffest section at the point of greatest transverse strain, which point, due to the inertia of the rod, is approximately 0.6 of its length from the piston end, it is found to produce very satisfactory results. Fig. 60 illustrates such a rod with the rod end in place. This rod is of the "strap" pattern, with wedge and screw for taking up the bearings at both ends. The adjustable feature is often omitted at the "wrist pin" bearing, but some method of taking up the wear should be provided for the crank-pin bushings. On the cheaper engines, and engines where lightness of parts is essential, the adjustable feature consists of metal liners of light-gage brass placed between the rod and the cap, which latter is held in place by studs and castellated lock nuts. As the bearing becomes worn, the cap is removed and a pair of liners taken out. Fig. 61 illustrates a rod commonly in use on engines of the marine and automobile type. It is known as the "I" section rod and is made of drop-forged steel, the web and flanges being formed in the dies. The cap and

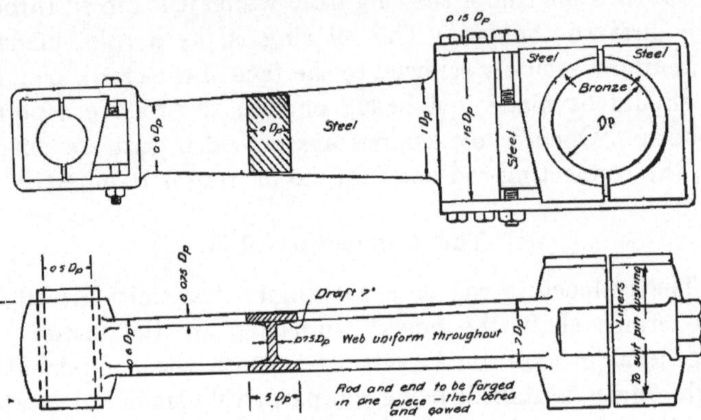


**Figs. 60 and 61. Connecting Rods.**

liners are shown in place. In both rods illustrated the general dimensions — based on good average practice — are given in terms of the crank-pin diameter. Kent gives the following simple formula for rectangular section connecting rods:

$$t = 0.01\, D \sqrt{P} + 0.6 \quad . \quad . \quad . \quad . \quad . \qquad (37)$$

In which $t$ = Thickness of rod in in.

D = Diameter of cylinder as given above.

$P$ = Maximum pressure in cylinder.

Having determined the value for $t$, the depth of the rod at the crank end is made equal to 2.25 $t$ and the depth of the wrist pin end 1.5 $t$. The length of connecting rod between centers varies between the ratios 2 : 1 and 3 : 1, in which 1 is equal to the length of stroke. The above formula is based on the required section at the surge point and, in consequence, surplus metal is found at the crank end.

## The Piston, Wrist Pin, and Piston Rings.

The piston of a gas engine is connected to the rod by means of the "wrist pin" bearing. There are two methods employed for forming this bearing. In the first — and the one most commonly in use — the wrist pin is locked fast in the piston, the rod working on it as a bearing. In the second method the pin is locked firmly in the rod, the bearing for the pin being in the piston itself. Fig. 62 illustrates the

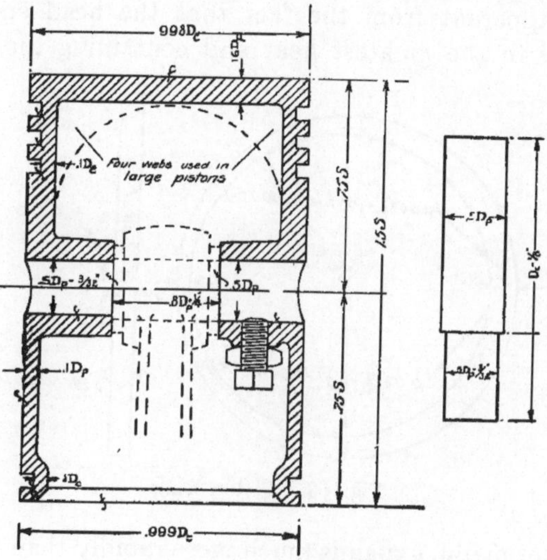


Fig. 62. Piston and Wrist Pin.

general form of piston in use for four-cycle engines. The weight of the piston, as well as the rod, should be as light as consistent with the strength required. The head and that part of the shell under the rings are made heavier than the remainder of the piston, as indicated in the drawing. The sizes, as indicated, are in terms of the crank-pin diameter and length of stroke. The connecting rod and bushing are shown in dotted lines, the wrist-pin bushing being pressed into the rod end and allowed to project, as shown, to secure the end bearing necessary. Finish marks indicate that part of the piston to be machined. The set screw and lock nut are shown in place. This screw should be made as tight a fit as possible in order to prevent its working loose. The wrist pin is shown beside the piston drawing, but while its diameter is given the same as the rod bushing, it should have a running fit of from 0.001 to 0.003 in., depending on its size. It should fit in the piston as tightly as may be without danger of cracking the shell.

The outer diameter of the piston should taper slightly from the head toward the crank end. The reason for this is very apparent from the fact that the head end, being subjected to the greatest heat and containing the greatest

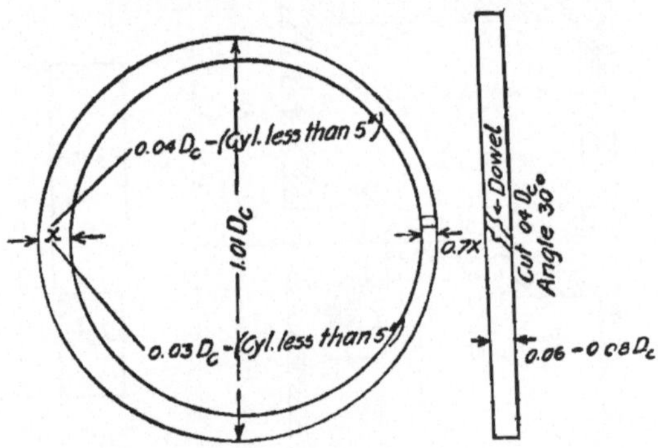


Fig. 63. Piston Ring.

amount of metal, expands much more rapidly than the crank end and is liable to stick and cut the cylinder. Obviously this taper should increase with the piston diameter. A good average rule is to make the head diameter equal to 0.998 $D_c$ and the crank end equal to 0.999 $D_c$. The length of the piston should approximate 1.5 the stroke with the wrist pin located practically at its center.

The best pistons are usually provided with four eccentric rings, although some of the cheaper ones are made with three or even two rings. The ring at the crank end serves in the capacity of an oil ring and also provides a bearing for that end of the piston. The piston ring is shown in Fig. 63 and its dimensions are given in terms of the cylinder diameter. The rings should be made of the best grade of gray cast iron and cast in a so-called "piston pot." A "piston pot" is generally made large enough to turn out eight or more rings; it is finished to the size required on the outside, and the inside turned to the proper eccentricity, after which the rings are cut off and slotted. The rings may then be chucked in an electric chuck and ground to the exact cylinder bore, or they may be used with the machine finish. The best engines have the rings ground, as this insures a ring perfectly round

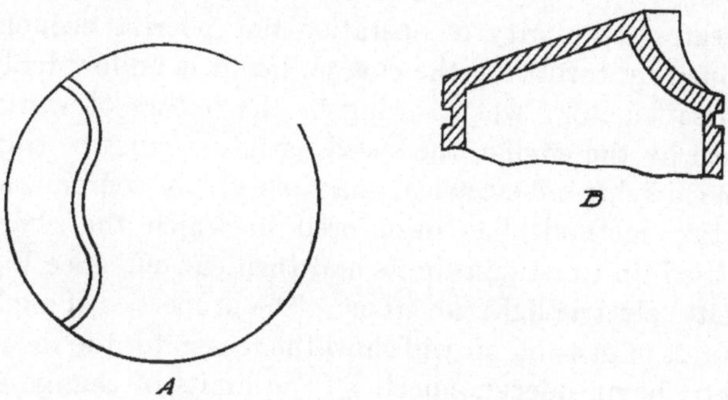


**Fig. 64. Two-Cycle Piston Head.**

when compressed into the cylinder. To maintain the rings in their fixed positions and with their joints properly staggered, a dowel pin is placed in the piston groove to correspond with the pin hole in the ring slot as shown. The rings are then fitted to their respective grooves, making them just small enough to contract and expand without binding.

Pistons for use in two-cycle engines are cast with suitable means for deflecting the incoming gases to the top of the cylinder. This may be accomplished by means of a web cast on the top of the piston, see Fig. 64, *A,* or in a more satisfactory manner by casting the piston as shown at Fig. 64, *B.* The general dimensions of the piston and rings remain the same as in four-cycle engines. The method of obtaining the exhaust-port lead in type B has already been discussed in the chapter on cylinders.

# CHAPTER XX.

## GOVERNING DEVICES.

THERE are four principal methods used for governing the speed of an internal-combustion engine, viz.: (1) By throttling the charge; (2) By cutting off the supply during one or more complete cycles, commonly known as hit or miss; (3) By keeping the exhaust valve open or closed during one or more strokes; (4) By interrupting the spark if electrical ignition is used.

Greatest regularity of operation but poorest economy is obtained by throttling the charge, but it is undoubtedly the most satisfactory when, owing to the nature of work performed by the engine, the speed variation must be slight as in the driving of electrical machinery. A combination of the two methods has been used in which the charge is throttled up to certain limits and then cut off. See Fig. 65, the Otto electric-light governor. The properties of explosive mixtures of gas and air will show the reason for the throttling method being uneconomical. "The limits of change allowable in the proportions of gaseous explosive mixtures are very narrow, the gas present ranging from $\frac{1}{7}$ to $\frac{1}{18}$ of the total volume. A mixture containing $\frac{1}{7}$ of its volume of coal gas in air has just sufficient oxygen to burn it and no more; any further increase of gas will pass away unburned, there being insufficient oxygen present for its combustion." — CLERK, "The Gas and Oil Engine," page 226.

A mixture of air and gas containing less than $\frac{1}{18}$ of its volume of gas loses its inflammability altogether. It follows that governors acting on the gas supply should be set to govern between these limits only.

The mixture itself may be throttled without altering the

ratio of gas to air, thus producing instead of an inferior mixture a greater or less amount of the same quality of mixture. This method reduces the force of explosion and at the same time maintains a uniformly explosive charge.

Keeping the exhaust valve open is of course wasteful of fuel and less economical, as is also stopping the spark. Keeping the exhaust valve closed produces a mixture of unburned products in the cylinder which must be completely exhausted before the cycle of operations can again be taken up.

The "hit or miss" type in which the fuel supply is entirely cut off for one or more complete cycles is of course the most

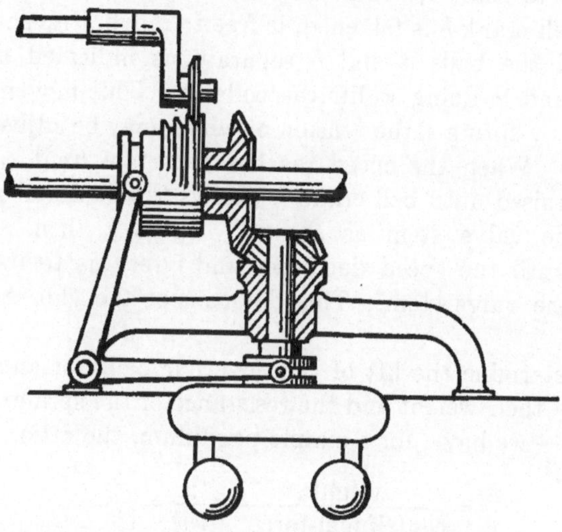

Fig. 65. The Otto Electric Light Governor.

economical method in use, as the charge is never varied and there is either an explosion of full force or none at all. It is apparent, however, that "hit or miss" governing admits of a greater range of speed variation, although it is possible with a properly adjusted governor to obtain regulation within four or five per cent as against two per cent with the throttling governor.

The mechanisms used to control the different governing devices are of two general designs, viz : (1) The centrifugal governor, which may be used for any one of the above-mentioned methods of governing; (2) The inertia governor which is applicable only to "hit or miss" governing. A modification of one of these two methods, more or less involved, is the principle of every governing device.

The centrifugal governor is essentially a revolving pendulum with inclined arms, advantage being taken of the tendency of these arms to revolve in the same plane as their points of support. Figs. 66, *A* and *B*, illustrate the principle of the centrifugal governor. At *A* the two fly balls *a* and *b* are connected, as shown, to collars *c* and *d*. Collar *c* is fastened to shaft *e*, while collar *d*, carrying the swivel *f* to which bell crank *h* is fastened, is free to move. As shaft *e* is revolved the balls *a* and *b* separate as indicated by the arrows and in doing so lift the collar *d*. This movement is resisted by spring *s*, the tension of which may be adjusted by collar *i*. When the speed reaches a certain limit collar *d* will be raised until bell crank *h* throws the push rod *j* away from the valve stem as shown. Valve *k* then remains closed until the speed decreases and *j* returns to its place below the valve stem. This device is of the "hit or miss" variety.

To determine the lift of the governor balls for any given speed — their weight and the resistance of the spring *s* being known — we have, for a simple pendulum, the ratio

$$\frac{h}{r} = \frac{\text{weight}}{\text{centrifugal force}} = \frac{w}{\frac{wv^2}{gr}} = \frac{gr}{v^2},$$

or

$$h = \frac{gr^2}{v^2}.$$

Substituting for $v$ its value $2\,\pi rn$,

$$h = \frac{9.788}{n^2}\text{ in.}$$

or

$$= \frac{35,237}{N^2}\text{ in.} \qquad (38)$$

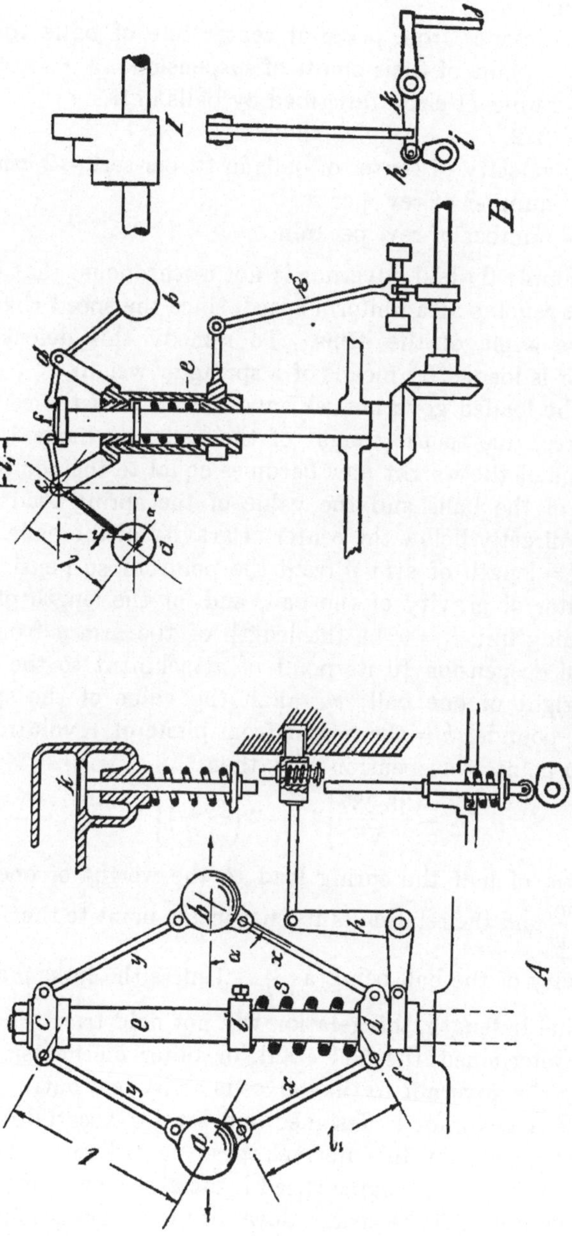

Fig. 66. The Centrifugal Governor.

In which

$h$ = distance from plane of center line of balls to the plane of their points of suspension.

$r$ = radius of circle described by balls in ft.

$g$ = 32.2.

$v$ = velocity of center of balls in ft. per sec. = $2\pi rn$.

$n$ = number of rev. per sec.

$N$ = number of rev. per min.

The simple fly-ball governor is not isochronous; that is, it does not revolve at a uniform speed, since the speed changes with the angle of the arms. To remedy this defect the governor is loaded by means of a spring or weight.

For the loaded governor we have the value of the centrifugal force, due to the weight of the balls, unchanged, but the value of the weight now becomes equal to the combined weight of the balls and the value of the spring load considered directly below the center of gravity of the balls.

Let $l$ = length of arm $y$ from the point of suspension to the center of gravity of the ball, and let the length of the suspending link $l_1 = x$ be the length of the arm $y$ from its point of suspension to its point of attachment to the ball; $w$ = weight of one ball; $w_s$ = half the value of the spring load in pounds; $h$ = the height from plane of revolution of balls to point of suspension of $y$; then

$$h = \frac{35{,}237}{N^2}\left[1 + 2\left(\frac{w_s}{w}\frac{l_1}{l}\right)\right] \quad . \quad . \quad . \quad . \quad (39)$$

the ratio of half the spring load to the weight of one ball being $\frac{w_s}{w}$ and the relation of its suspension point to the center of gravity of the ball being as $\frac{l_1}{l}$. Unless the links $y$ and $x$ are equal in length, this relation will not hold true. The lift being determined, the bell crank or other mechanism connecting the governor to the valves is easily laid out.

Since a governor is designed and set for a certain speed it must be run at this normal speed regardless of that of the engine. If the engine speed is increased or diminished, the governor must be geared down or up as the case may be.

At $B$ is shown another application of the centrifugal ball type of governor. The lift of the balls, $a$ and $b$, may be determined by means of another application of the ratio

$$\frac{h}{r} = \frac{\text{Weight}}{\text{Cent. Force}},$$

$$\frac{h}{r} = \frac{W - W_s \dfrac{l}{l_1 \sin \alpha}}{\dfrac{Wv^2}{gr}}$$ (Considering the spring load to be combined with the weight of the balls.)

Simplifying, $$h = \frac{0.815}{n^2}\left((1 + \frac{W_s l}{W l_1 \sin\alpha}\right) \quad . \quad . \quad . \quad (40)$$

In which $n$ = The rev. per sec.
$W_s$ = Half the spring load.
$W$ = The weight of one ball.
$\alpha$ = The angle of the arm $l$.

If we allow $N$ to represent the rev. per min., then the formula will be $h = \frac{2934}{N^2}\left(1 - \frac{W_s l}{W l_1 \sin\alpha}\right)$.

This formula for small movements of the balls is sufficiently accurate for all purposes. In this style of governor, referring to the figure, we have two levers acting on $f$ as a fulcrum and exerting their lifting force at $c$ and $d$. The bell crank $g$ rides the cam roller $h$ off or onto the cam $i$ as the speed varies, and valve $j$, actuated by lever $k$, opens or remains closed as the case may be. It would be possible, by making the lift of the cam variable as shown at $I$, to produce a throttling effect at the valve. Throttling with a poppet valve, however, produces an abnormal amount of wire drawing of the charge. A better arrangement is to use an independent throttle valve in the gas main. Fig. 67 represents such a throttling valve with governor connection. As the balls $a$ and $b$ move outward the bell crank $c$ moves with the collar $d$ and the valve $V$ is shoved in the direction of the arrow $e$. The openings $f$ in the valve gradually move across the openings $g$ in the shell $h$, reducing the effective port openings into the annular space $i$, and throttling the mixture.

The principle of the inertia governor is illustrated in Fig. 68. The weight $a$ is connected to bell crank $b$ and pivoted to the push rod $c$. The tendency of the weight to turn the bell crank in the direction of the arrow is

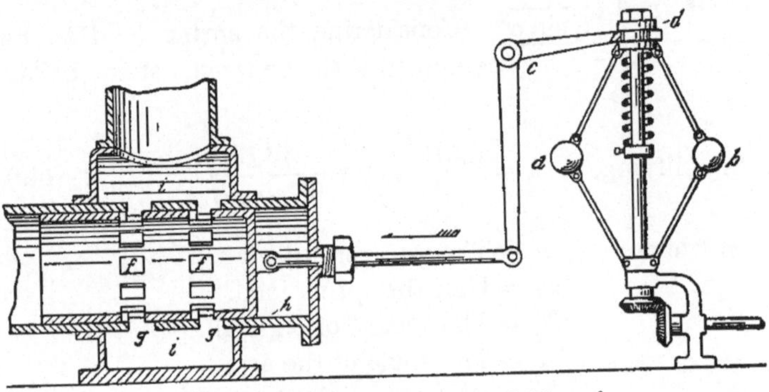

Fig. 67. Centrifugal Governor and Throttle.

resisted by the spring $d$, but as the push rod is moved rapidly the inertia of the weight partially overcomes this resistance and the engaging parts of the valve gear are separated when the speed reaches a certain limit. The strength of the spring required is determined by means of the following formula:

$$W = \frac{W_a l + \frac{WV}{g} l}{l_1} \quad . \quad . \quad . \quad . \quad . \quad (41)$$

In which

$W$ = Capacity of the spring in lb.
$W_a$ = Weight of $a$ in lb.
$l_1$ = Lever arm of $a$ about $c$.
$V$ = Velocity attained by push rod $b$ in ft. per sec.
$g$ = 32.2 ft. per sec.

In the above formula the weight of the rod carrying the pendulum is neglected. To be accurate the resultant moment of it and the weight should be used, but by threading the weight onto the rod, as shown, adjustment may be secured by moving it in or out as required. Inertia governors are used, as a rule, on industrial engines where regularity of operation is a secondary consideration to the cost.

Electrical governing is accomplished by switching off the current for the ignition system by adapting some one of the above-mentioned governing devices to the requirements. When the engine attains a certain speed the current is cut off, and for one or more cycles the engine runs without an impulse. Electrical governing is of the "hit or miss" type.

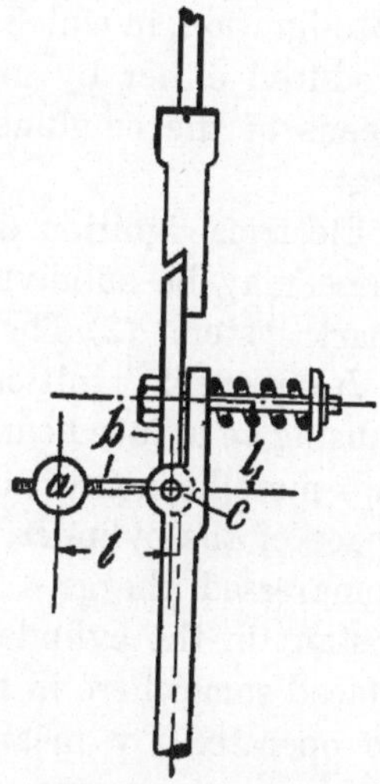


Fig. 68. Inertia Governor.

In order to operate well a governor mechanism must be as sensitive to the variations of speed as it is possible to make it. For this reason the practice of placing the mechanism in the crank case is unfortunate. The device, when in this location, is constantly subjected to the splash from the crank shaft, and the dirty oil soon gums and causes the parts to stick. The engine will then be found to be governing poorly, and the result in electrical installations is a fluctuation in the e.m.f. If it is deemed advisable to inclose the governing mechanism, a separate chamber should be used for this purpose and the governor inspected and cleaned from time to time.

# CHAPTER XXI.

## IGNITION.

THERE are three general methods employed for securing the ignition of the compressed charge in a gas-engine cylinder: (1) By means of an electrical spark; (2) By means of a mechanically operated flame or heated surface; (3) By auto-ignition, in which heat sufficient to ignite the charge is produced either by means of the compression alone or by means of the combined effect of compression and residual heat.

Electrical ignition devices are most extensively used, and these may be subdivided into two classes: (1) The jump-spark system; (2) The make-and-break system.

Jump-spark ignition, as the name implies, consists in causing an induced current of high potential to spark between two metallic points conveniently placed in the compression space of the cylinder, the spark thus produced igniting the compressed charge. To produce the spark at the proper instant in the cylinder a make-and-break contact must be placed somewhere in the electrical circuit, this contact being so operated by means of the crank or cam shaft that at some point in every operating cycle the circuit will be closed and an e.m.f. generated sufficient in value to cause a spark to jump across the gap. This make-and-break device is generally known as the commutator or spark advancer, the latter name having its origin from the fact that the commutator is so arranged that the engine operator, by means of a suitable lever, is able to change the point at which the charge is ignited so that it may correspond to the speed at which the engine is running. This statement may be somewhat confusing inasmuch as the speed regulation is obtained, to a very large extent, by advancing or retarding the time of ignition. Nevertheless the point in the cycle must correspond to the speed at which the engine is operating, at any particular instant, in order to secure smooth running. In starting the engine the spark is set at a point slightly beyond dead center and *in the expansion stroke.* As soon as the explosions commence to occur regularly the spark is gradually advanced past dead center and *into the compression stroke,* thus giving the engine time to gather speed. If the spark is rapidly thrown over dead center into the compression stroke the engine will either stop or pound badly until it attains sufficient speed to carry itself from the point of ignition up to dead center before the burning gases reach their maximum pressure.

The make-and-break system of electrical ignition consists in causing, by mechanical means, two points or electrodes located in the compression space to close and then open the electrical circuit. This may be accomplished by causing the points to rub together and then separate, producing what is known as the "wipe spark," or by forcing the points together and causing them to separate by means of a spring, sometimes called the "hammer break." When the points break contact an intensely hot spark or arc is produced, due to the inertia of the electric circuit producing, momentarily, a very high potential.

The make-and-break spark is much hotter than the jump spark and on reasonably slow speed engines is the most satisfactory form of electrical ignition, it being almost certain that, if the points are in good condition, a spark hot enough to ignite the charge will be produced at every contact. On the other hand, the points are subject to much wear, especially with the wipe spark, and consequently deteriorate quite rapidly. If platinum alloy is used for the points they are usually quite expensive, and the necessity of replacing them is troublesome as well. Within the past few years other alloys, which it is claimed by their manufacturers give better satisfaction, have been placed on the market. "Casalloy" or "meteor-wire" is one of these substitutes.

The points of a make-and-break ignition may, with proper care and a current of proper strength, be made to last a long time. If the electric pressure is too low, unsatisfactory ignition will result, while, on the other hand, should the pressure be too high the plugs will require adjustment or renewal in a very short time. With a battery of low internal resistance, as a storage cell, the difference of potential at the terminals should be much less than with cells of high internal resistance. In order that a sufficiently large number of primary cells may be carried for all emergencies and to allow for

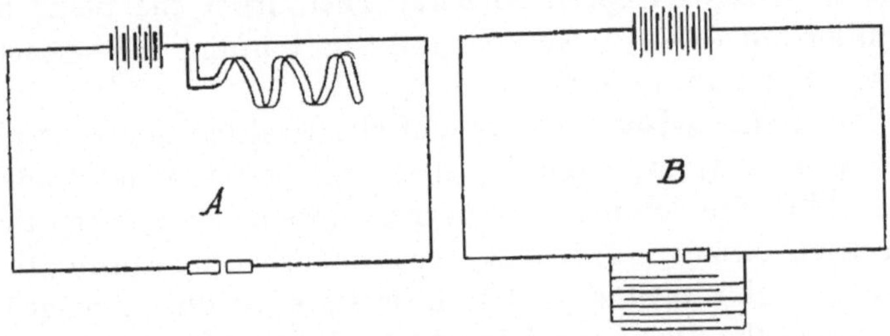


Fig. 69. The Non-Inductive Resistance and the Condenser.

depletion, the destructive action at the points may be reduced by placing a condenser in parallel with the points or by introduction into the circuit of a non-inductive resistance. Fig. 69 illustrates clearly these two methods. A non-inductive resistance thrown into the circuit in series causes a fall in potential without producing any unbalanced electromagnetic action which would affect the sparking coil. Such a resistance coil is made by doubling a wire, placing the closed end on a bobbin of wood or other non-magnetizable substance, and winding the wire about the bobbin as indicated at *A*, Fig. 69. The electromagnetic action in one wire is thus neutralized by an exactly similar and equal action in the other. It is apparent that if this method is used a suitable resistance box with varying resistances to accommodate itself to varying conditions must be provided. If the condenser method is used and a condenser large enough to meet all requirements is provided, no further adjustment or attention will be necessary.

The ordinary commercial condenser consists of sheets of tin foil as shown in *B*, Fig. 69, insulated from one another with the alternate sheets connected to terminals. The current for the ignition is obtained by means of primary cells, storage batteries, or a small dynamo inducing high potential secondary current in an induction coil.

For the primary-secondary jump-spark ignition system a secondary current of momentarily extremely high potential

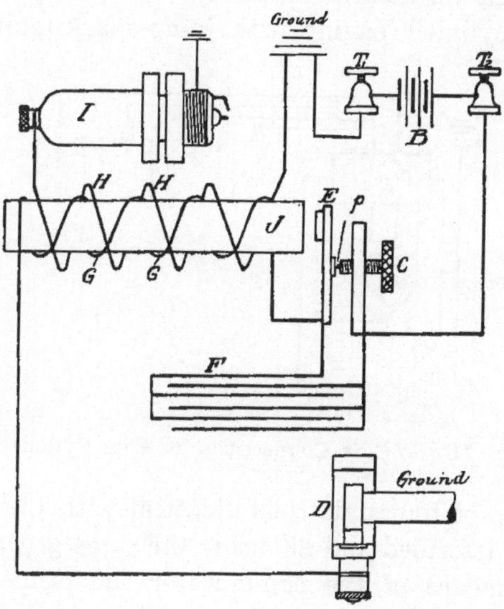


**Fig. 70. The Ruhmkorff Coil and Connections for Single Cylinder.**

is required; this is obtained by means of the so-called Ruhmkorff induction coil or vibrator. Fig. 70 shows the construction of this coil as applied to a single cylinder.

$T_1$ and $T_2$ are the battery terminals of coil of which *T* is the ground terminal; *B* the battery; *C* the contact screw with platinum point *p*; D the commutator, one terminal being grounded by means of its own shaft closing the primary circuit with *T* when the commutator points are in contact; *E* the vibrator of soft iron which opens and closes the circuit at point *p*; *F* the condenser, which momentarily arrests the primary current and minimizes the break spark at $p$ (the condenser is not essential and many coils are made and operate successfully without it); $G$ the primary coil of heavy wire carrying battery current; $H$ the secondary coil of heavily insulated fine wire; $I$ the spark plug, to one terminal of which the secondary coil is connected, the other terminal being connected through the ground to the opposite end of the secondary coil, and $J$ the iron core of very soft annealed wire.

Fig. 71 shows, diagrammatically, the wiring connections for a four-cylinder engine with jump-spark ignition. The

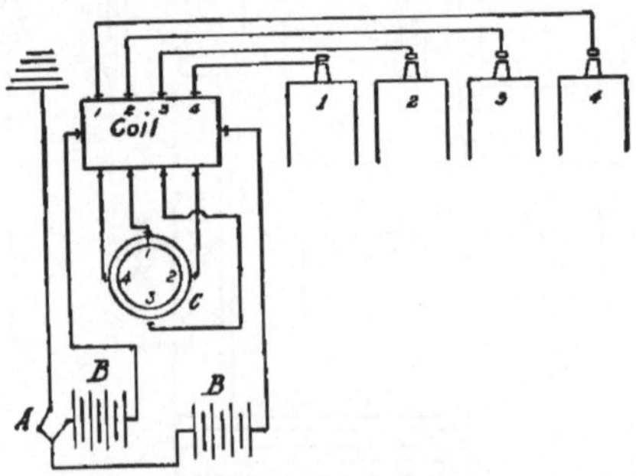


Fig. 71. Wiring Connections for Four Cylinders.

coil consists of four unit coils identical with the single coil previously described and all using the same ground connection. By means of the commutator the primary current is made to pass alternately through the units 1, 2, 3, and 4, producing in each, in turn, a high-potential current which fires the charge in the cylinder to which it is connected. The system as shown is wired for two sets of batteries, $B$ and $B_1$. By means of the three-point switch $A$ either set of coils may be thrown into the primary circuit, so that if one is found to produce an insufficient spark the other set may be used.

The coil is a very important element in the successful operation of a jump-spark ignition system. Numerous makes of coils are marketed, many of which are apparently very reasonable in cost, but it is doubtful if it is advisable, in any case, to purchase a coil and consider its low cost as a recommendation. The coils which give best satisfaction are rather expensive, but it will be found in the long run that they are most economical. To give good results, especially on high-speed engines, a coil must be fast, that is, it must charge and discharge very rapidly. In order to accomplish this result the core must be of best annealed iron wire and the coils well insulated, especially the high-tension windings. With a sluggish hungry coil the best of engines will give poor results.

Every coil is rated to give a certain length of spark, but the points of the spark plug must be set very much closer than this maximum, since it requires much greater pressure to spark across a gap under high compression than in a vacuum or at atmospheric pressure. The strength of current must also govern the spark gap. For average practice $\frac{1}{16}$ in. is about right for this gap.

The vibrator coil is not suitable for a make-and-break ignition system. The vibrating high-potential current will cause the ignition points to spark as they come together, or a spark may be produced at some equally undesirable point in the cycle. The ordinary single induction coil, or "booster," produces the best results, unless a sparking dynamo is used, in which case no coil is necessary, the current from the dynamo being sufficient to produce a good spark. In using the make-and-break system of ignition no commutator is used, as the current is interrupted at the ignition points and the timing of the ignition is secured by changing the point in the cycle at which they separate.

In the jump-spark ignition system the spark is produced when the circuit is closed, this result being accomplished by the vibrator interrupting the current and producing the necessary high potential many times a second, while in the make-and-break system, this high potential being produced only once, as the points separate, it is at that point that the spark occurs. Fig. 72 illustrates a make-and-break system in which the points $p$ and $p_1$ are forced apart suddenly without the wiping effect. In the figure, $A$ is a quick-return cam which in the position shown is about to trip push rod *B*. On this push rod is located the stop and adjustment collar *C* which impinges against the fiber washer D when the circuit is broken. The flat compensating spring *E* takes up the motion of the push rod *B* after the points *p* and $p_1$ are in contact. Coil spring *F*, acting

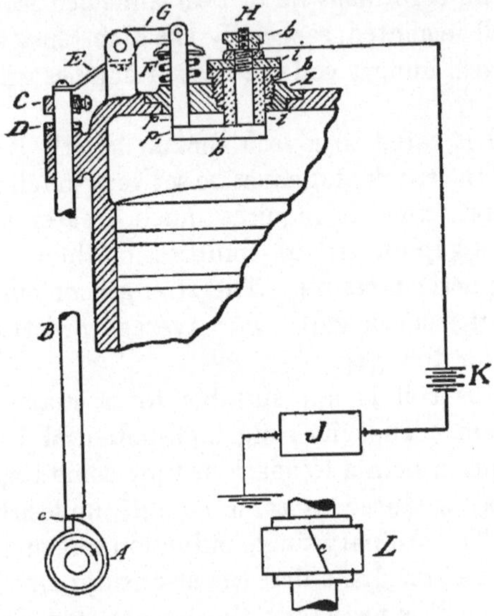


Fig. 72. **Hammer-Break Igniter.**

on lever *G*, as shown, throws the sparking points sharply apart as the point of the cam *A* leaves the push rod. The plug *H*, which carries the battery terminal $p_1$, must be constructed of porcelain, or other insulating material equally as good, and brass. In the plug as shown the parts marked *b* are brass and those marked *i* are insulation. The entire sparking device is set into the cylinder head on the plate *I* and held in place with cap screws. The wiring connections are exceedingly simple, the coil *J* and battery *K* being connected up in series. The free terminal of the coil is grounded on the engine frame and the free terminal of the battery connected to the plug *H* as shown. It is apparent that the time of ignition may be varied by causing cam *A* to trip early or late, thus advancing or retarding the spark.

This may be accomplished by making the edge *c* of the cam as shown at *L*, Fig. 72. Then by moving the shaft in the direction of *X* the spark may be advanced, the push rod tripping sooner than if the shaft were shifted toward *Y*, in which position the spark would be retarded.

Fig. 73 shows a water-cooled hammer-break igniter, similar to that in use on the Rathbun gas engines.

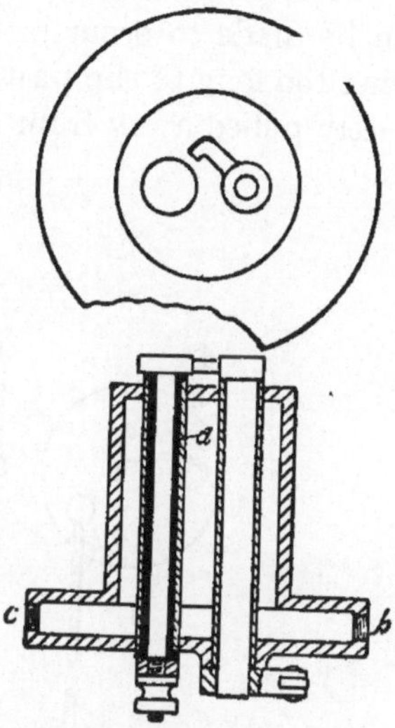

Fig. 73. Water-Cooled Igniter.

Fig. 74 shows a common type of wipe-spark ignition mechanism. The point *p* mounted on the oscillating device *A* makes and breaks contact with the spring point *p*. This oscillating movement is produced by means of the cam and push rod as shown, the return movement being obtained by means of spring *B*. Timing may be accomplished by changing the tripping point of the cam as described. The ignition points should be located as nearly in the path of the incoming gases as possible in order to secure the greatest cooling effect. The electrode points in this igniter are not made of platinum alloy, and for that reason this form is inexpensive and, if properly designed, efficient.

Ignition mechanisms attached to the piston head are used to some extent. The Pennington igniter is one of these and its principle is illustrated in Fig. 75. The circuit-breaking device *A* is tapped into the piston head, as shown, the insulated terminal being tapped into the cylinder and consisting of a coiled spring carried on the forked end of plunger *C*, as shown. As the piston *P* nears the end of its stroke the stirrup-shaped end of *A*, which has an inclined surface as shown, strikes *B*. If plunger *C* has been pulled out, as indicated by the arrow, only the point of *B* will be engaged by *A* and only a slight movement of the piston will be required to cause it to slip off of *A* and produce a spark. Under these conditions the ignition will be advanced. If, however, plunger *C* is pushed farther in, the contact break can be made to occur in the explosion stroke, the spring *B* being too long to slip past the inclined position of *A* which is merely pulled away from *B* as the piston recedes. With the

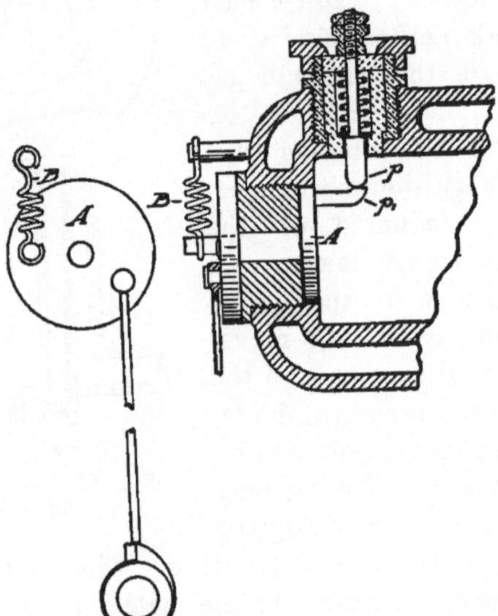


**Fig. 74. Wipe-Spark Igniter.**

ignition advanced it is apparent that this mechanism will produce two sparks, one as the finger *B* slips off *A* into the stirrup and another as it snaps out on the return stroke of the piston.

The commutator or spark advancer, as used in connection with the jump-spark ignition system, is made in a variety of different forms and a description of one typical form should be sufficient for all purposes. For four-cycle engines it is either mounted on the cam shaft or geared to it with 1 : 1 gearing, it being apparent that, since but one explosion occurs in each complete cycle, the spark as well as the valve mechanism must be operated by a half-time shaft. In two-cycle engines the commutator is either mounted on the crank shaft or geared directly to it without any reduction in time.

A commutator consists essentially of a piece of fiber, or other tough insulating material, carrying one terminal, and another terminal, connected to the ground, which once in every cycle comes in contact with the insulated terminal, thus momentarily closing the electrical circuit.

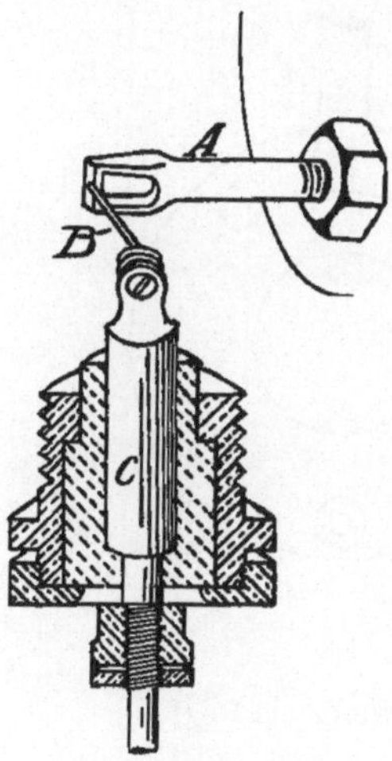


Fig. 75. The Pennington Igniter.

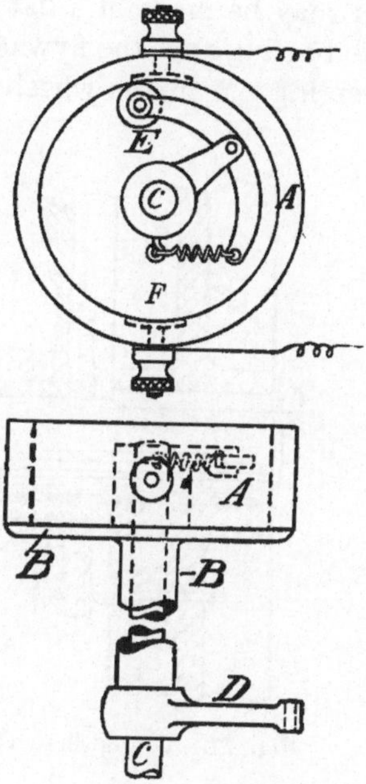


Fig. 76. Commutator for Two-Cylinder Engine.

Fig. 76 shows a common form of commutator for a two-cylinder engine. In the drawing, *A* is a fiber ring mounted as shown on a flanged sleeve *B*, which may be rotated about shaft *C* by means of the lever D. Shaft *C*, which may be the cam shaft or an auxiliary timing shaft, carries the ground contact mechanism *E*. As the shaft *C* revolves, and the contact *E* engages the shoe *F*, the circuit is closed through the primary coil, and the current from the battery causes an electromotive force to be generated in the secondary winding, producing a spark. It is apparent that by shifting the sleeve *B* around the shaft *C* the ignition may be advanced or retarded by varying the point where the ground contact *E* meets the shoe *F*. A commutator similar to the one described may be made for any number of cylinders by increasing the number of contacts accordingly.

For two-cycle engines a simple and inexpensive commutator may be made of a flat fiber disk mounted on the end of the bearing next the flywheel, the ground contact point being a spring pin in the wheel. Fig. 77 represents this form of

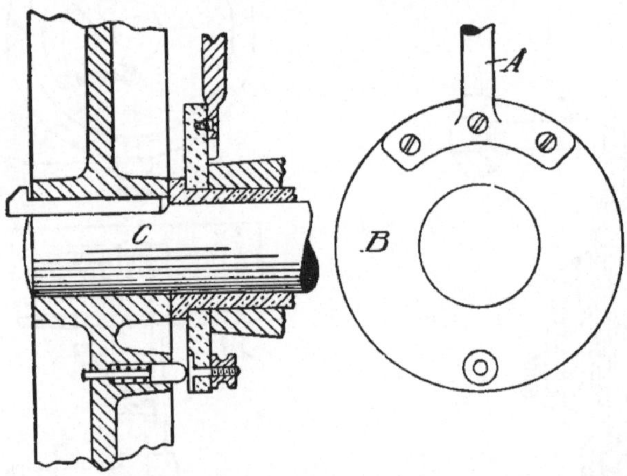


Fig. 77. Crank Shaft Commutator for Two-Cycle Engine.

commutator. Means for advancing or retarding the spark is provided by lever *A*, by which the fiber may be moved about the shaft *C*.

Fig. 78 represents a form of commutator, of cheap construction, which may be used to advantage in single-cylinder four-cycle construction. The make and break is secured by means of the cam *C*, mounted on the cam shaft, acting on the spring *S* which is carried on the fiber *F* as shown. Timing is secured by moving the fiber around the shaft by means of the rod *R*.

Fig. 79 illustrates a more complicated and expensive commutator.

Commutators are on the market which distribute to the secondary while timing the primary current, thus making it possible to use a single coil to spark multiple cylinders. While this method reduces the number of unit coils and vibrators necessary, the system is not as good as one using the multiple units in which the commutator current is low tension and much more easily insulated.

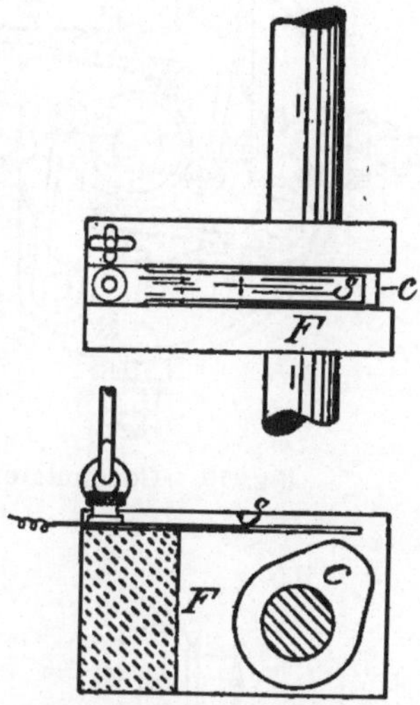

Fig. 78. Cam Shaft Timer for Four-Cycle Engine.

Spark plugs for use with jump-spark ignition systems consist of a porcelain or lava insulated point projecting into the explosion chamber and in close proximity with another similar point connected to the ground through the cylinder. Many forms are offered for sale.

By far the most important point in a spark plug is to make sure that there is perfect insulation, otherwise under the high compression the spark will jump across at some point in the circuit where the resistance is less than at the spark gap. While this may not be apparent with the plug withdrawn from the cylinder, it is often the cause of serious ignition trouble.

Fig. 80 shows a spark plug of French make which is an efficient but expensive plug. In the illustration, *A* is the grounded and *B* the insulated point between which the spark jumps across the gap *C*. The porcelain insulation is in three pieces, as shown, with suitable packing, where the brass parts bear upon it, to prevent breaking. The plug must necessarily have its joints gas tight in order to prevent leakage of the compression, and the several joints make this style of plug particularly efficient in this respect.

Cheaper plugs are made that give excellent satisfaction. The author has had the best results with the so-called

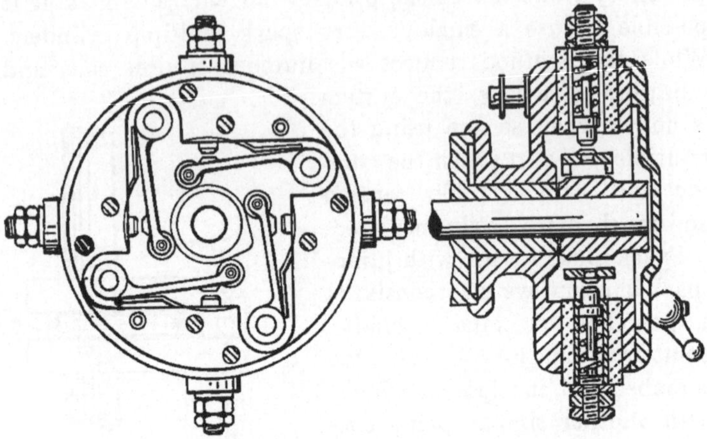

Fig. 79. Commutator with Hammer Break Contacts.

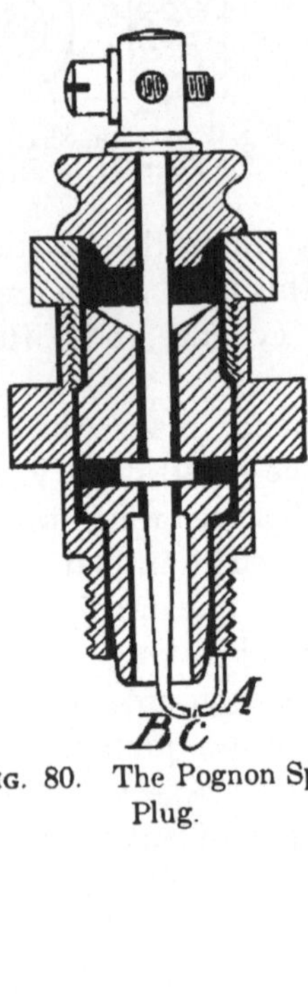


Fig. 80. The Pognon Spark Plug.

Fig. 81. Spark Plug of Ordinary Construction.

"Rajah" plugs, although the porcelains being light they are subject to breakage. Figs. 81, 82, 83, and 84 represent several well-known makes of plugs.

Dynamo ignition has become quite popular in the past few years, and were it not for its rather high first cost it would undoubtedly be the universal electrical ignition. It is absolute in its operation, and used in connection with a

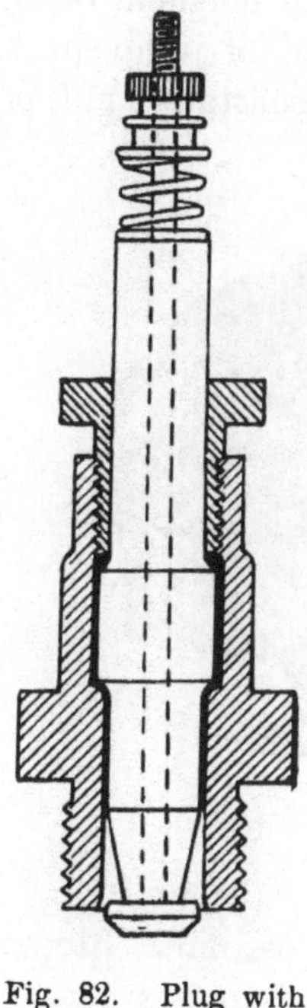

Fig. 82. Plug with Spring Clip Connection.

Fig. 83. Sta-Rite Plug.

Fig. 84. Sta-Rite Plug.

storage cell, the operator need give no attention to the matter of batteries. Without the storage cell it becomes necessary to use two or three cells for starting, if the engine is too large to turn over quite rapidly by hand, after which, by means of the dynamo, it generates its own sparking current.

Of the dynamo igniters probably the one most commonly in use is the Apple made by the Dayton Electrical Manufacturing Company, Dayton, Ohio. It is a neat and compact device, the reasonable first cost of which, as well as its efficiency, recommends it. The parts are entirely enclosed in a water and dust proof case. It is provided with a centrifugal friction clutch governor, as shown in Fig. 85, the shoes of which release as the engine speed increases, thus causing it to run steadily. The brushes are so placed as to enable the dynamo to run in either direction, the field magnets being permanent. These igniters are made either for jump-spark or make-and-break ignition, and generate a constant e.m.f. of

Fig. 85. The Apple Igniter.

from 4 to 5 volts at from 1000 to 1200 rev. per min. Storage batteries may be connected up to these dynamos and the surplus energy not used for sparking stored up for starting.

Fig. 86 shows the wiring diagram for a single cylinder with jump-spark ignition. For a multiple cylinder the system is identical except that there are extra spark plugs and commutator connections to be made. See Fig. 87. The storage battery, Fig. 88, always furnishes the current to the primary, and by means of the four-point automatic cut-out switch $S$, see Fig. 89, it is possible to read on the volt-ammeter the voltage of the storage battery at any instant, the amount of current being put into the battery by the dynamo, and the current each coil consumes. Fig. 90 shows the wiring diagram for make-and-break ignition in which the double coil is not used. Fig. 91 shows the Apple igniter in section.

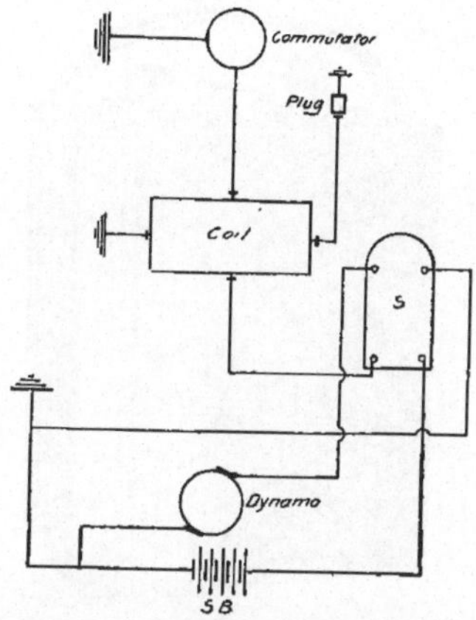


Fig. 86. Wiring Diagram for Jump Spark with Apple Igniter.

In Fig. 92 is illustrated a dynamo of the Bosch type. The stationary armature *A* is provided with two windings *a* and *b*, of which *a* corresponds to the primary winding of an induction coil and *b* corresponds to the secondary winding. The current is produced by the rotation of a soft-iron sleeve *B* which is integral with the hollow shaft *C* and which partially surrounds the armature *A*. Shaft *C* also carries the notched disk D and the high-tension distributor *E*.

Fig. 87. Wiring Diagram for Four Cylinders.

One end of the primary winding *a* is grounded on the shaft, and the secondary *b* forms a continuation of the primary. The opposite end of the primary *a* is led to one side of contact breaker *F* and also to one terminal of the condenser, the other terminal of which and the moving arm of contact breaker *F* are grounded. Sleeve *B* is slotted, and when the slots come opposite the poles of the field magnet *G*, the armature

Fig. 88. Storage Battery.

Fig. 89. Four-Point Cut-Out.

windings momentarily cutting the magnetic lines, a powerful e.m.f. is generated. The contacts of the contact breaker *F* are held together by the action of the disk D during the interval between sparks; the low-tension winding *a* thus being closed on itself, there is a high-tension current through

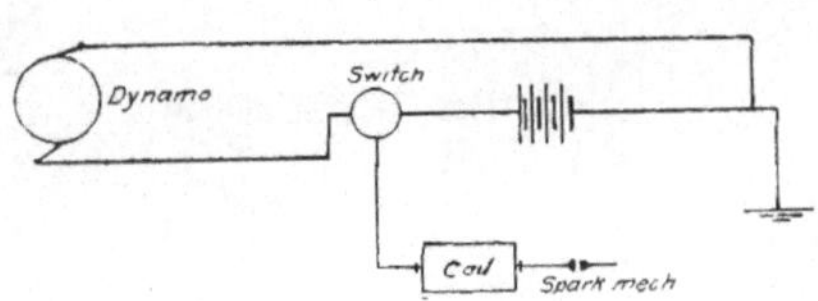


Fig. 90. Wiring Connections for Make-and-Break Ignition.

it as the magnetism of its core is varied by the rotating sleeve *B*. When one of the notches shown on *B* comes under the movable arm of the contact breaker *F* the points snap open, owing to the action of the spring, and break the primary circuit *a*. This generates a high e.m.f. in the secondary winding *b*, the *current* being further increased by means of the condenser *H*. The secondary winding being connected to the primary as described and connected through it and the ground to the several spark plugs as shown, sparks are produced in the four or more cylinders, the tension of both the primary and secondary coils being utilized.

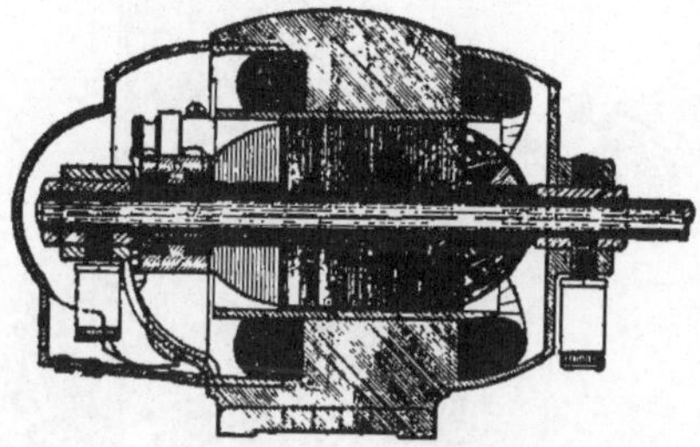

Fig. 91. Section through Apple Igniter.

Distribution is accomplished by means of the commutator *E* which opens and closes the secondary coil.

The Motsinger Auto-sparker is shown in Fig. 93. It

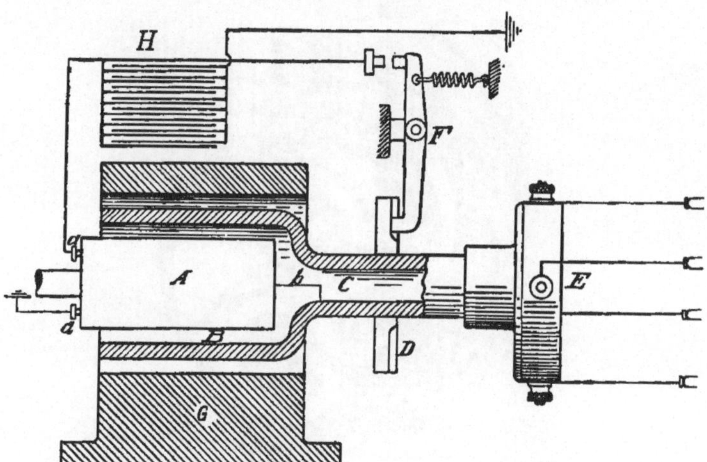


Fig. 92. Bosch Type of Dynamo.

runs at from 700 to 1200 rev. per min. and is automatically governed to maintain a uniform speed.

A cut of the Remy Magneto, made by the Remy Electric Company at Anderson, Ind., is shown in Fig. 94. This igniter is so designed as to produce two electrical impulses for each revolution of its armature, and must consequently

Fig. 93. The Motsinger Sparker.

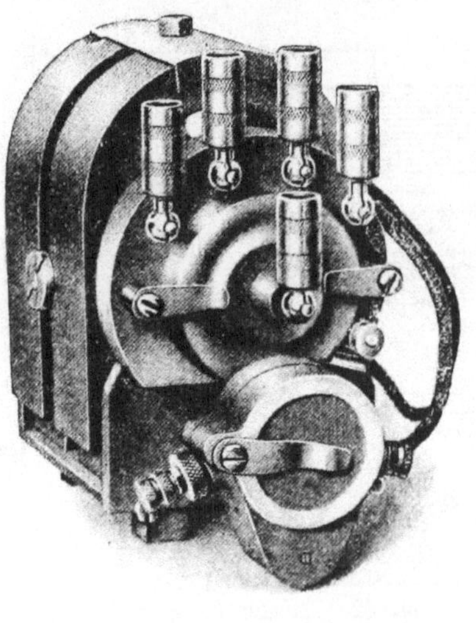

Fig. 94. The Remy Magneto.

be in step with the engine. It will deliver a spark at from 25 to 30 rev. per min. and it is easy to start an engine with this magneto by cranking it in the ordinary way.

Flame or hot-tube igniters are still used to some extent on engines for general purposes, and give good uniform results. Notwithstanding this they are being very rapidly replaced by electrical ignition systems. The simplest of the hot-tube igniters is the direct-flame type in which a gas burner outside the cylinder is brought into contact with the charge by means of a port being uncovered, at the proper instant in the stroke, by the piston movement or by valve gear on the crank shaft. A complicated form of the flame igniter is found on the Otto slide-valve engine. Barnett's ignition cock, Fig. 95, illustrates the general principle of the flame igniter.

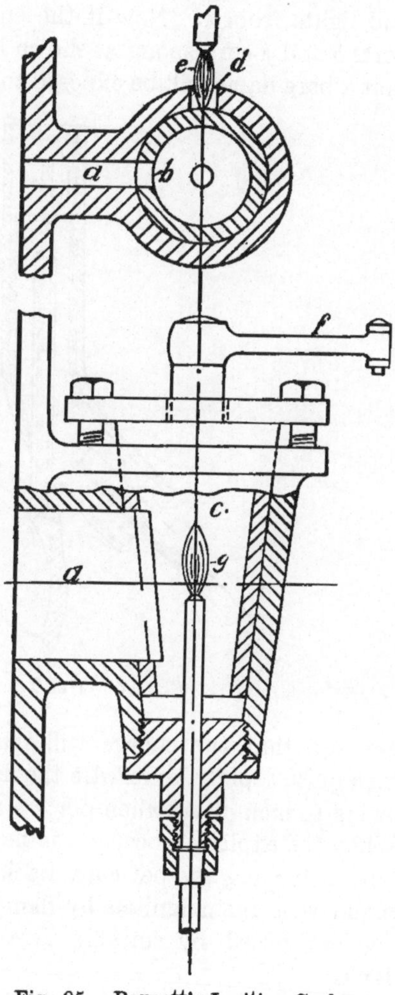

Fig. 95. Barnett's Ignition Cock.

In the illustration, *a* is the ignition port opening to the explosion chamber from *b*, the port opening in *c*, a hollow pet cock which may be turned so that *b* will be opposite to port *d*, a port which is open to the atmosphere and in front of which *e*, a gas flame, is kept burning; *f* is a lever connected by means of some suitable half-time mechanism to the crank shaft and by which port *b* may be made to alternately open to ports *a* and *d*, and *g* is a gas jet contained in the hollow pet cock *c* as shown.

In starting, the engine is turned over until ports *b* and *d* correspond and gas jet *e* is lighted. Jet *g* is then turned on and lights from *c*. Now if the engine is turned over until ports *b* and *a* correspond as shown in the figure it is apparent that a bare flame will be exposed to the contents of the cylin-

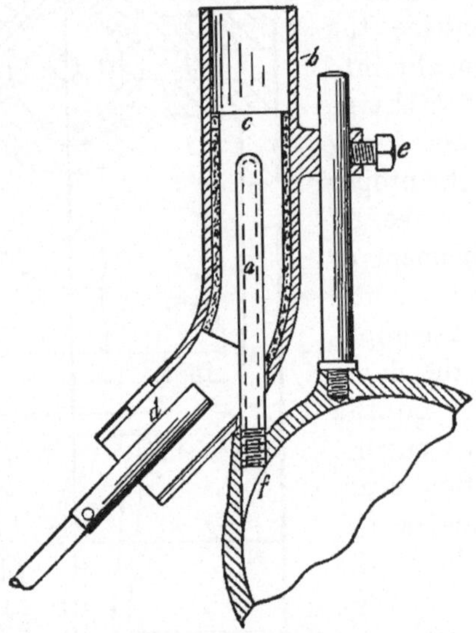


Fig. 96. Hot Tube Igniter.

der and that the charge will ignite. The pet cock must turn quite rapidly, otherwise the flame *g* will be extinguished owing to lack of air when ports *a* and *b* are both closed to *b*. When the explosion occurs the flame *g* will be extinguished, but on turning the pet cock back until ports *b* and *d* correspond it is again ignited by flame *e*. The time of ignition may be varied by suitable arrangement of the half-time device.

The hot-tube ignition is well illustrated by Fig. 96 and is similar to those in use on the Robey engines.

In the figure, *a* is the igniter tube, which should be of nickel alloy or porcelain; *b*, the chimney with lining of asbestos;

*c*, the asbestos lining; *d*, the bunsen burner which heats *a*, and *d*, a set screw by means of which tube *b* may be raised or lowered to regulate the degree of heat to which *a* shall be raised.

On the compression stroke of the engine the charge is forced into tube *a*, through ignition port *f*, and at the proper instant, determined by the compression, is exploded. The adjustment of flame and tube to secure the proper degree of heat, that the charge may be ignited at the proper instant, is a delicate operation and one that is determined only by experiment for the different makes of engines. It is to be remembered that the products of combustion tend to remain for some time in the ignition tube and that the next charge of gas must force these products ahead of it before reaching the hot part of the tube.

The hot-tube igniter, in connection with a timing valve by which the ignition may be more readily varied, is also used to some extent. The tube, in this style of igniter, is shorter and of larger bore, and is kept at a temperature high enough to insure the charge igniting as soon as it comes in contact with the heated surface. The ignition tube of a hot-tube igniter should be of some substance not easily affected by the severe heat to which it is continually subjected. Nickel alloy has been found to give the best satisfaction, in metal tubes, a tube of this material giving service for a long time, unless subjected to the action of sulphurous gas or excessive heat, while a tube of wrought-iron pipe must be replaced every few days. Porcelain tubes, although more difficult to set, give excellent satisfaction for long periods. The R. Thomas & Sons Company, East Liverpool, Ohio, furnish porcelain tubes; and rods for the making of nickel-alloy tubes may be secured from H. Boker & Co., 101 Duane Street, New York.

Fig. 97 shows a good method for setting porcelain tubes in a socket that may be screwed into the cylinder. The tube may be packed in with asbestos washers and moist fire clay or with asbestos cement.

Auto-ignition, or ignition secured from the heat of compression, is found, in more or less modified form, in several makes of engines. Chief among these is the Diesel motor. In this motor the compression is carried to about 500 lb. per sq. in. above atmosphere, and the temperature in the combustion chamber reaches nearly 1000 deg. fahr., which is sufficient to ignite the charge, as soon as it is forced into it. The ignition of this engine as well as its fuel valve are fully described elsewhere in this work, under their respective heads.

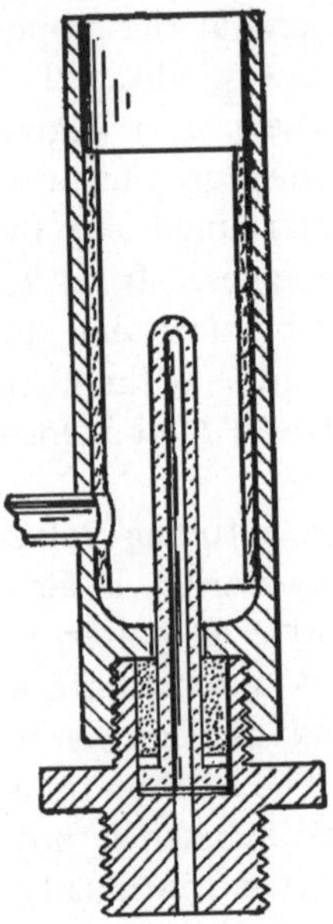

Fig. 97. Method of Inserting Porcelain Tube.

A modified form of this type of ignition, and one in which the pressure is not carried as high, is found in the Hornsby-Akroid and the Mietz & Weiss oil engines, the former being of the four-cycle type and the latter a two-cycle engine. Fig. 98, *A* and *B*, shows the principle of these two igniters. In both, ignition is obtained by forcing the charge, during the compression cycle, into a small chamber *a*, the walls of which are not water-jacketed but are surrounded by an outer casing *e*. The heat of compression, together with the residual heat from previous explosions, in this small insulated chamber is sufficient to ignite the charge. In *A*, the four-cycle type, the oil is sprayed directly into the igniter, thus utilizing the heat in this chamber to vaporize the oil. The same mechanism, operating the oil spray *b*, opens the air port *c* on the suction stroke, at the end of which stroke the igniter chamber contains vaporized oil and products of combustion, while the cylinder contains almost pure air. On the compression stroke, the air being forced into chamber *a*, the temperature, due to compression, rises until at the proper instant ignition occurs. On the exhaust stroke, the exhaust valve *d* is opened and the products of combustion are forced out of the cylinder.

These products will, due to the isolated location of the valve, remain to quite a large extent in the ignition chamber and probably produce a slight back pressure at the end of the exhaust stroke, due to their inability to pass through the restricted neck leading to the cylinder as rapidly as necessary. The residual products quite naturally produce a

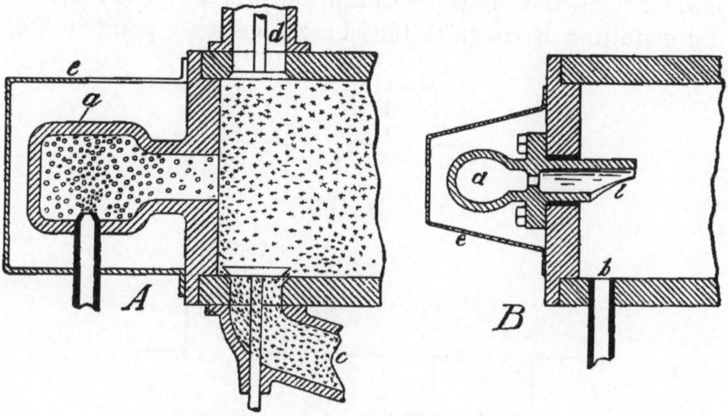


Fig. 98. Types of Auto-Ignition.

slightly inferior mixture, but results of tests show the operation of these engines to be quite satisfactory.

The two-cycle type, illustrated at *B*, sprays the fuel oil into the cylinder at *b* and against the heated lip *l*, and while it does not have the advantages of the vaporizing effect secured by impinging the oil against the heated surface of *a*, the vaporizing effect being secured by means of the lip *l*, on the other hand it does not mix intimately with the products of combustion contained in *a*; with the result that better mixture with the air is secured. This engine works on the principle of the two-cycle three-port engine, with the exception that the fuel is not drawn into the crank case but is injected into the partially compressed air as it enters the cylinder from the crank case. It is apparent that this type of ignition would not be applicable to engines depending on the igniter for their speed variation, except as the fuel charge might possibly be throttled so as to decrease the compression and make the explosion occur late. The speed variation is obtained by varying the fuel charge.

There is no fixed method of determining at what point in the cycle ignition should occur; with different fuels and under varying conditions this point will be different with nearly every case. Gasoline vapor ignites and burns very rapidly; gaseous fuels consume more time in the process. The theoretical condition is to ignite the charge at such a point in the

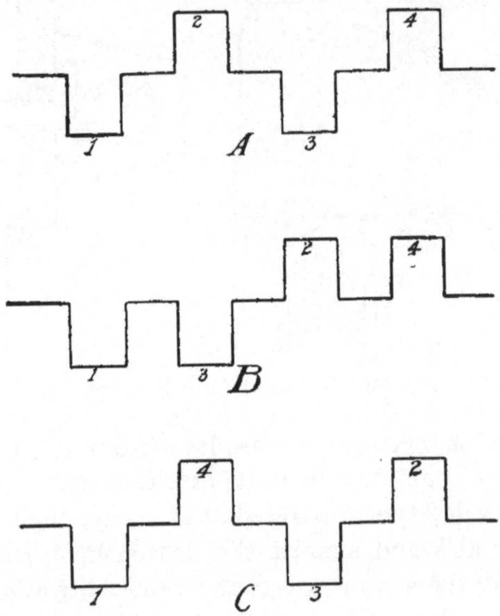


Fig. 99. Four-Throw Cranks for Different Firing Orders.

compression stroke that it will attain its maximum pressure just as the engine passes dead center into the expansion stroke. Approximately it may be said that the lead should be increased 5° on the crank circle with every 100 rev. per min. increase in speed; thus at 200 rev. per min. the lead should be about 10°, at 300 rev. per min. about 15°, and so forth. The point at which the engine runs best should be determined by opening the throttle wide and then advancing the spark until with further advancement no greater speed is obtained or the engine commences to knock.

The firing order of engines with more than two cylinders varies to some extent with the manufacturers. A three-cylinder engine should fire in order $\overset{1}{1}-\overset{2}{2}-\overset{3}{3}$,* in fact it can be fired no other way, the cylinders following each other from left to right or from right to left. A four-cylinder engine may be fired (*A*) $\overset{1}{1}-\overset{2}{2}-\overset{3}{3}-\overset{4}{4}$, or (*B*) $\overset{1}{1}-\overset{2}{3}-\overset{3}{2}-\overset{4}{4}$, or (*C*) $\overset{1}{1}-\overset{2}{3}-\overset{3}{4}-\overset{4}{2}$, the cranks, of course, being made to correspond. The order may be taken from either the left or right. For firing order *A* the crank throws must be as shown at *A*, Fig. 99. For firing order *B* the crank throws must be as shown at *B*, Fig. 99, and for firing order *C* the crank throws must be as shown at *C*, Fig. 99. It is apparent that the cranks for firing orders *A* and *C* are identical, the wiring connections only being changed. Order *B* is probably the best practice and as widely in use as any.

In a six-cylinder four-cycle engine the explosions overlap, one impulse being given every third of a revolution. The order of firing should be $\overset{1}{1}-\overset{2}{3}-\overset{3}{5}-\overset{4}{2}-\overset{5}{4}-\overset{6}{6}$, with cranks set 60° apart. For an eight-cylinder engine the order of firing is $\overset{1}{1}-\overset{2}{5}-\overset{3}{3}-\overset{4}{7}-\overset{5}{2}-\overset{6}{6}-\overset{7}{4}-\overset{8}{8}$, with the cranks set 45° apart; that is, crank number 5 is located 45° behind number 1, crank number 3 is located 45° behind number 5, and so forth.

* In the above firing orders the small figures represent the cylinders, calling the one to the left 1. The larger figures below indicate the order in which that particular cylinder is fired.

# CHAPTER XXII.

## ENGINE TESTING.

THERE are but two general methods of testing engines to determine their b. hp., although there are many different designs and makes of instruments used in the running and recording of these tests. The two methods are (1) By absorption of the power; (2) By transmission of the power. The instruments used in the first case are known as absorption dynamometers and those used in the second case as transmission dynamometers. The absorption dynamometer differs from the transmission dynamometer in that its object is to absorb the delivered power at its point of delivery and to record its force and velocity; while the transmission dynamometer measures the actual net differential tension of the belting or gearing and, with its velocity known, determines the amount of power delivered by it. Transmission dynamometers are of two kinds — the self-recording and those in which the readings must be constantly taken by the operator from a dial and the mean pressure obtained from a number of these readings.

Fig. 100 represents the simplest form of absorption dynamometer or Prony brake. In the figure, $A$ is the radius of the drum or pulley to which the friction is applied as shown, the bolts $C$ and $C$ clamping the device as tightly as may be necessary. The lever arm $B$ delivers the absorbed force to the spring balance $E$, where from time to time, at regular intervals, readings are taken by one operator while another, by means of a tachometer, determines at what speed the pulley of radius $A$ is revolving. Now it is apparent that if the end of lever arm $B$ were unrestricted it would continue to revolve as a part of the pulley $A$ and that the point $X$, at which it is shown attached to the balance $E$, would travel at some speed directly proportional to the number of revolutions being made by $A$. At each revolution it would describe a circle of radius $R$. The circumference of the circle traveled over by the point $X$ would then be $2\pi R$. Now suppose a resistance just sufficient to hold the lever arm $B$ stationary were applied at $X$. This resistance, in terms of weight, then would be the force exerted by that point throughout its complete revolution. We would then have $F = 2\pi RW$ (for one revolution). Now knowing the revolutions per minute being made by pulley $A$ at the time

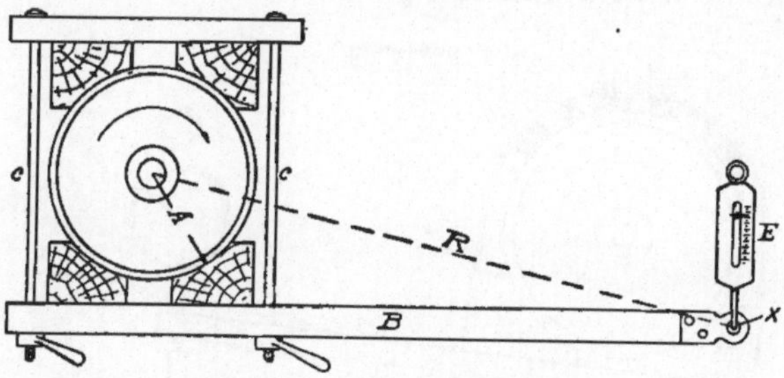


Fig. 100. The Prony Brake.

the reading $W$ is taken, we may determine the force that would be exerted in one minute at the point $X$ or $F = 2\pi RWN$ ($R$ being in feet and $W$ in pounds the result is clearly in foot-pounds). Now since the foot-pound equivalent of a horsepower is 33,000 foot-pounds per minute, the horsepower being delivered at $X$ may be found by dividing the above by 33,000, or

$$H = \frac{2\pi RWN}{33{,}000} \quad \cdots \cdots \quad (42)$$

*Design of a Prony Brake.* — In a shop where a number of engines of the same design are turned out and it is desirable to test each engine before it leaves the shop, a specially designed brake wheel is made. The probable horsepower, $H$, is known from previous calculations, also the speed in rev. per min. at which the engine is to run. Call the peripheral speed of the flywheel $V$ and let $W$ be the width of the brake in inches. R. H. Thurston, in "Engine and Boiler

Trials," tells us that the number obtained by multiplying the width of brake $W$ by the peripheral speed $V$ and dividing by the horsepower $H$ shall not exceed 500 to 1000.* Calling this value $K$, we have the formula

$$K = \frac{WV}{H} \quad . \quad . \quad . \quad . \quad . \quad . \quad . \quad . \quad (43)$$

From which the unknown value $K$ may be readily found.

In running a brake test on a gas or gasoline engine, especially the latter, where the speed is high, difficulty is encountered in the heat produced by the friction of the brake on

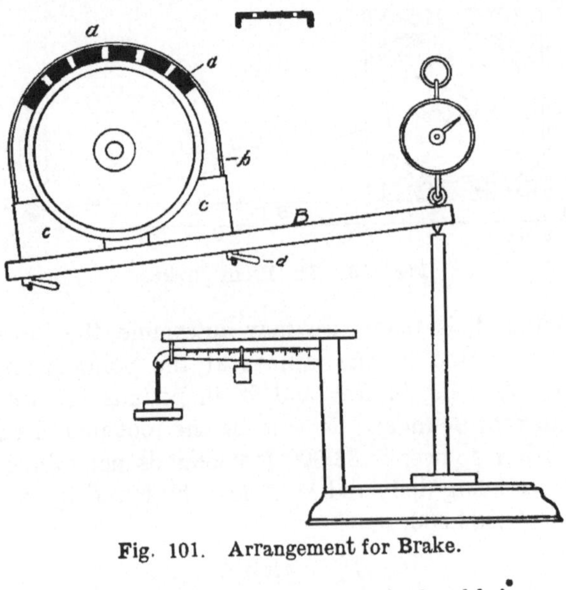


**Fig. 101. Arrangement for Brake.**

the wheel. For this reason a method should be provided for supplying water to the rim. One method is to make the flywheel rim in the shape of an inverted U, pouring in water as the engine is running. The centrifugal force carries the water out in the rim and keeps it there. This method is satisfactory for comparatively short tests, but for longer running tests a more efficient device should be used. A good way is to so design the wheel that running

* "Engine and Boiler Trials," R. H. Thurston, pages 272 to 279.

water may be kept flowing into it. Fig. 101 illustrates the design of such a wheel. The construction of the brake is as follows: the small blocks *a*, made of fiber or tough wood, are securely riveted to band *b* which passes around the wheel and through blocks *c* which are fastened to the lever arm *B*. Pressure of the brake on the wheel is obtained by means of the clamping nut *d*. The wheel has a rim section, as shown, through which 0.25-in. holes are bored at random. Water may now be kept running in a steady stream into the rim and out through these holes onto the brake blocks. As the

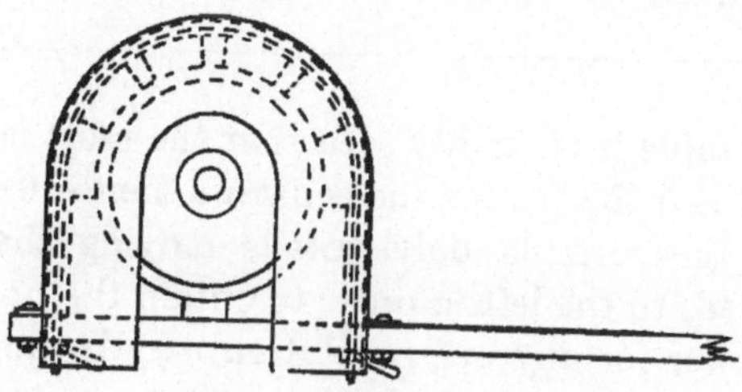

**Fig. 102. Brake Shield.**

rapidly revolving wheel throws the water and makes the testing an exceedingly unpleasant operation, a shield of sheet metal should be placed over the brake, as shown in Fig. 102. The addition of a little soap is a good precaution to prevent the brake sticking and causing jerky and uneven action of the scale beam.

*Discussion of the Formula.* — The formula for the Prony brake, as above derived, admits in all cases of a constant being determined which simplifies the solution. There are in the formula four known quantities: $\pi = 3.1416$, $R =$ the radius of the lever arm as designed in any particular brake, 33,000, the foot-pound equivalent, and the figure 2. Suppose now that the brake were designed with a lever arm $R = 1$ ft., then $H = \frac{2 \times 3.1416 \times 1 \times WN}{33,000}$. Simplifying the formula gives us $H = 0.019\, WN$. It is apparent that for any given length of brake arm a similar constant may be derived as shown in Table XIII.

TABLE XIII.

PRONY BRAKE FACTORS.

| Length of Brake Arm in Feet. | Factor by which to Multiply $WN$ to give $H$. | Length of Brake Arm in Feet. | Factor by which to Multiply $WN$ to give $H$. |
|---|---|---|---|
| 1 | 0.019 | 6 | 0.114 |
| 2 | 0.038 | 7 | 0.133 |
| 3 | 0.057 | 8 | 0.152 |
| 4 | 0.076 | 9 | 0.172 |
| 5 | 0.090 | 10 | 0.190 |
| 5.252 | 0.100 | 11 | 0.200 |

From the table it is readily seen that the most convenient length for $R$ is 5.252 ft., for the constant then is 0.1 and the solution of the formula only entails moving the decimal point one place to the left in order to obtain the horsepower. The horsepower for any one revolution may be obtained by simply multiplying the scale reading at any instant by the constant.

The brakes most in practical use are modifications of the ones described, but there are some specially designed brakes which obtain the friction by hydraulic action, the water producing the pressure acting as a cooling agent as well. The Alden brake is so constructed. For a more exhaustive discussion of the Prony brake the author would refer the reader to Carpenter's "Experimental Engineering," pages 207 to 216.

*The Belt Dynamometer.* — A modification of the Prony brake which may be used in the testing of small engines is the belt dynamometer, in which the brake arm is the radius of the pulley or flywheel. The belt to be used is provided at both ends with a ring, to which weights may be fastened, as shown in Fig. 103. The acting force, analogous to $W$ in the formula, in this case is equal to the difference between the weights $A$ and $B$ when the apparatus is in equilibrium.

Then $$H = \frac{2\,\pi R\,(A - B)\,N}{33{,}000} \quad . \quad . \quad . \quad . \quad (44)$$

An improvement of the above form of dynamometer is obtained by the substitution of spring balances for the weights, the balances being placed under initial tension as shown in Fig. 104. The formula remains the same.

To run a complete test on an engine, the brake constitutes but a small part of the outfit required. The operator

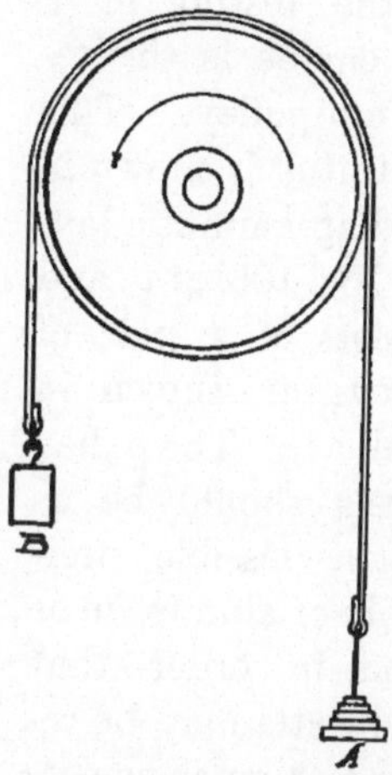


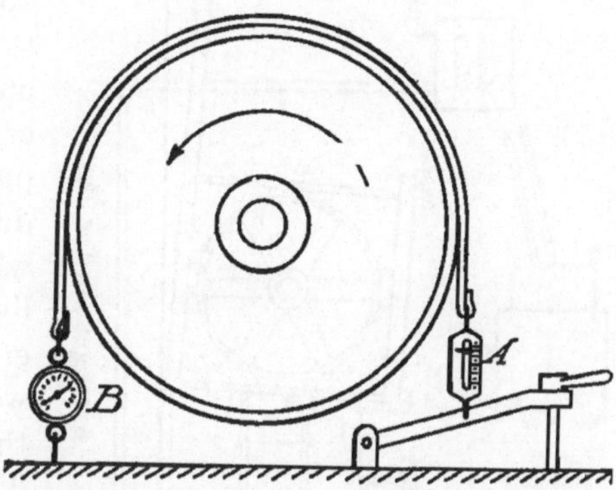


Fig. 103. Strap Brake. Fig. 104. Strap Brake with Spring Balances.

must have, besides, a good indicator with a stiff spring suitable for gas-engine work, several standard thermometers, a recording speed indicator or tachometer, an air pyrometer or other suitable instrument for taking the temperature of the exhaust, an auxiliary scale for weighing the discharge water during any minute of the test, a reliable barometer; a gas meter is generally connected to the engine and so need not be supplied, but in order to obtain the ratio of the air supplied to the fuel consumed an air meter should be placed on the suction. The gas pressure should be obtained, if required, by placing a water pressure gage in the gas supply and obtaining the value in inches of water. Since the gas pressure is subject to some variation, for extreme accuracy a rubber gas bag should be placed in the supply pipe to reduce this fluctuation to a minimum. Before starting the test sufficient time must be taken to arrange the appara-

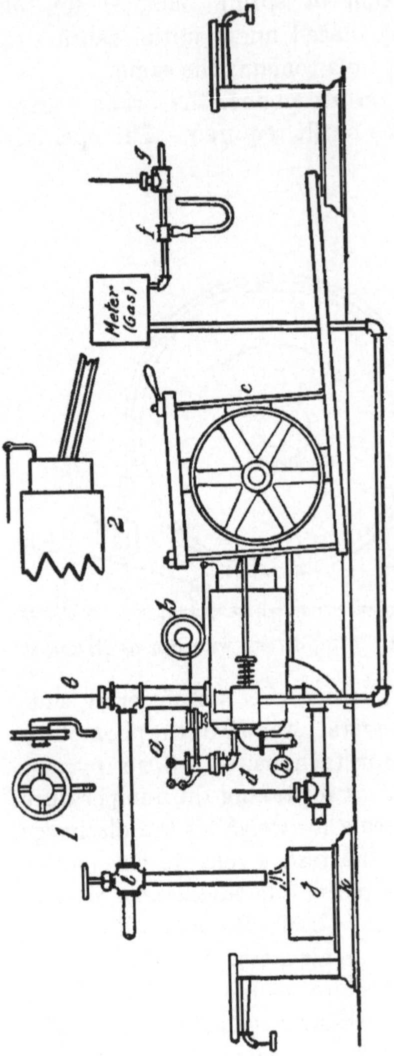


Fig. 105. Arrangement for Test.

tus. A reducing device, consisting of pulleys or a pantograph ("lazy tongs") must be rigged up. The device best adapted to the testing of a gas engine is the reducing pulley. This apparatus is shown in the diagrammatic layout, Fig. 105, at *b*, and consists of a pair of pulleys as shown in detail at *l*. The pulley wheels should be as light as possible, preferably of aluminum or wood, in order that the inertia may be reduced as much as possible. The device is tapped into the top of the water jacket, as shown, and two or three wraps of strong cord are taken around both pulleys, the end of the cord being made fast to the rim of the wheel. The cord from the larger pulley is then run to the piston end and fastened there rigidly, as shown at 2; the cord from the smaller pulley is then fastened to the cord on the indicator drum, as shown. For engines operating at a high rate of speed the indicator drum spring must be set up to high tension in order to absorb the inertia of the parts as rapidly as possible. The cord must also be free from metal hooks in order to minimize the tendency to vibrate. For an engine running at 200 revolutions per minute the cord tension should be about 2 pounds. For 300 revolutions per minute it should be 4.5 pounds; for 400 revolutions, 8.0 pounds, and for 500 revolutions, 14 pounds. The Prony brake *C* is set up substantially as diagramed. A thermometer in a brass oil bath is inserted in the discharge, as shown at *e*, and another in a similar oil bath is placed in the inlet, as shown at *d*. The temperature and pressure of the gas entering the engine are taken at *g* and *f* respectively. The pyrometer is located in the exhaust passage, as at *h*, in order to obtain the temperature of the exhaust gases. A three-way valve is placed at *i* so that the discharge water may be turned on or off the weighing tank *j* as required. The barometer and room thermometer may be placed at any convenient point near the test. The air meter is not shown but should be placed at some point in the air suction pipe.

*Testing with the Prony Brake.* — In running a test with the preceding arrangement, the first consideration is the weight which the brake itself will impose on the scale, a constant which must be subtracted from all scale readings in order to find the net value of *W* or the actual force being exerted on the scale by the brake arm. To determine this constant loosen the brake on the wheel or pulley and by means of a spring balance attached to the point at which the weight reading is taken, raise the arm a foot or more, reading the scale while doing so. Then lower the arm to its original position, taking another reading during the operation. As there is some friction on the wheel at any time, these two readings will be different, the first being the larger. The average of the two is the value of the constant *C*.

Constant *C* having been determined, prepare a log sheet for the test substantially as follows.

NAME OF ENGINE.........................................

MANUFACTURER .........................................

TEST MADE BY.........................................

AT.........................................

DATE .........................................

DIAM. PISTON.... ...STROKE .......CLEARANCE.......

| Number of Reading. | Time. | R.P.M. | Explosions per Min. | Temp. Room. | Temp. Exhaust. | Temp. Inlet Water. | Temp. Disc. Water. | Wt. Disc. Water. | Scale Read. Pounds. | Net Scale Read. | Cu. Ft. Gas. | Temp. Gas. | Press. Gas. Inches Water. | Cu. Ft. Air. | Barometer. | M.E.P. | I.H.P. | B.H.P. | Mech. Eff. $\frac{B.H.P.}{I.H.P.}$ |
|---|---|---|---|---|---|---|---|---|---|---|---|---|---|---|---|---|---|---|---|
| | | | | | | | | | | | | | | | | | | | |

Before commencing the actual test two or three assistants should be secured in order that the different readings may be taken as nearly at the same time as possible. One reading of the barometer and the room thermometer is generally sufficient for an hour's run. The engineer in charge of the test should handle the speed indicator while an assistant takes the brake scale readings, keeping the scale beam constantly floating as the power fluctuates. Another assistant will be able to handle the indicator and to take the temperature of the inlet and discharge water, while a fourth assistant will be able to take care of the weighing tank and to take whatever other readings are necessary. Readings should be taken at five-minute intervals while the test lasts, and an hour's test is generally sufficient for any one adjustment of the engine. One run should be made at maximum power, one at the rated horsepower, and a third at no load with the brake removed, the latter being merely an indicator test. Runs may also be made at quarter, half, and three-quarter load if desired. In order that the readings may be taken as nearly as possible at the same time the engineer in charge should be provided with a whistle, and signal ten seconds before a reading is taken and again as it is to be started. A stop watch is valuable, although not absolutely necessary, to the person handling the speed indicator. After completing the test, the clearance of the engine, its stroke and piston diameter should be carefully measured and recorded.

To measure the clearance, place the engine on its upper dead center, being sure that the valves are carefully seated. Now weigh a quantity of water, more than sufficient to fill the compression space; then fill the compression space from this measured quantity, being careful not to spill any part of the water, and weigh the remaining water. The difference between the two weights $W$ and $W_1$ is the weight of the water contained in the compression space of the engine. Now a cubic inch of water at 39.1 deg. fahr. weighs 0.036 lb., hence the weight of the water contained in the compression space divided by 0.036 will give accurately the number of cubic inches in that space. If the temperature of the water is much higher than 39.1 deg. and extreme accuracy is required, its weight per cu. in. may be determined by means of the following thermodynamic formula:

$$W = \frac{.036 \times 2}{\dfrac{t + 461}{500} + \dfrac{500}{t + 461}} \quad . \quad . \quad . \quad . \quad . \quad (45)$$

Wherein $t$ = temperature of the water.

In taking a series of tests with a transmission dynamometer the same general method as to readings is adhered to. There are so very many different forms of transmission dynamometers that the author, since the scope of this work is necessarily limited, would refer the reader to standard engineering works embracing the subject.* It is sufficient to say that the transmission dynamometer is placed between the prime mover and the driven machinery — the power transmitted being measured by the tendency to rotate gears, which tendency is resisted by springs of known resistance,

* See Carpenter's "Experimental Engineering," pages 219–234, Thurston's "Engine and Boiler Trials," page 264, or Weisbach's "Mechanics," Vol. II, pages 39–73.

or by weights, suitable recording arrangements, either automatic or dial, being provided.

The method of engine testing as described applies to the making of a complete test of a gas engine, in every detail. Such a complete test is not usually run except in cases where a record of the actual performance of an engine is required in every such detail as a basis for a guarantee or to detect faulty design. A description of a complete test naturally takes into consideration all the smaller details, but it is an easy matter to perform any part of the test required. The most common tests performed are the brake and indicator tests to determine the horsepower and the mechanical efficiency of the engine.

*Gasoline, Alcohol, and Oil Engines.* — In the testing of engines operating on liquid fuel, while the indicator and brake tests, as well as a majority of the other tests, will be run in the same way, the tests for fuel consumption will be different. In place of the column "cu. ft. of gas," a column reading "gal. of fuel" should be substituted. The column "press. of gas in in. of water" is omitted, and the column reading the "temperature of the gas" should be replaced by one reading the "temperature of the fuel"; in all other respects the log sheet may be used as it stands. The temperature of the fuel, if a carburettor is used, should be taken at the carburettor; if a jet or mixing valve is used the temperature should be taken just before the fuel reaches the valve.

In a gasoline automobile motor, while many manufacturers make talking points of their low fuel consumption, the question of the amount consumed is secondary to the power derived, flexibility of control, and weight, and, as a matter of fact, with the varying loads and speeds to which such an engine is constantly subjected it is impossible to obtain any very great fuel economy, and as a consequence a brake and indicator test is all that is usually required for such an engine.

## CHAPTER XXIII.

### REPORT OF TESTS.

THE data having been obtained by the method discussed in the last chapter, a report of the conditions found must be made.

When a number of tests are being conducted a printed form of report blank should be made as follows:

PLACE.................. .DATE......................

ENGINEER IN CHARGE............ ...........

ASSISTANTS.............. ...... .. .. ...... ....

NAME OF ENGINE. ...... ...MANUFACTURER...............

RATED HP.....,.............REV. PER MIN............ .

DIMENSIONS.

| | | |
|---|---|---|
| Stroke..................................... | —— | ft. |
| Diam. piston.............................. | —— | in. |
| Area piston................................. | —— | sq. in. |
| Piston displacement....................... | —— | cu. in. |
| Compression space........................ | —— | cu. in. |
| Compression space........................ | —— | per cent |
| Theoretical compression................. | —— | lb. |

DATA.

| | | |
|---|---|---|
| Length of test.............................. | Average | —— |
| Rev. per min................................ | " | —— |
| Rev. per hr.................................. | " | —— |
| Explosions per min........................ | " | —— |
| Explosions per hr.......................... | " | —— |
| Gas per hr.................................... | " | —— |
| Air per hr..................................... | " | —— |
| Ratio gas to air............................. | As 1: $x$ | —— |
| Water consumption........................ | lb. | —— |
| Temperature inlet water.................. | deg. fahr. | —— |
| Temperature dis. water.................... | " " | —— |
| Range water temperature.............. | Av. " " | —— |
| Temperature room......................... | " " " | —— |
| Temperature exhaust...................... | " " " | —— |
| Pressure inlet water....................... | Av. lb. | —— |

PRONY BRAKE.

| | |
|---|---|
| Length lever arm ($R$) | —— ft. |
| Constant brake (Correction) | —— lb. |
| Brake load average (Gross) | —— lb. |
| Brake load average (Net) | —— lb. |
| Weight of gas per cu. ft. | —— lb. |
| Weight of air per cu. ft. | —— lb. |
| Average mixture | Ratio gas to air —— |
| Weight of mixture per cu. ft. | —— lb. |
| Sp. heat gas | —— |
| Sp. heat air | —— |
| Sp. heat mixture | —— |
| Heat value cu. ft. | —— B.t.u. |

RESULTS.

| | |
|---|---|
| Work | Av. ft.-lb. per min. —— |
| Work | Av. ft.-lb. per hr. —— |
| d.hp. | Average —— |
| Indicated m.e.p. | Av. lb. —— |
| Indicated hp. | Average —— |
| Gas per i.hp. | cu. ft. —— |
| Gas per d.hp. | cu. ft. —— |
| Mech. efficiency | Av. d.hp. ÷ Av. i.hp. —— |
| Friction loss | i.hp. − d.hp. —— |

HEAT PER HOUR.

| | |
|---|---|
| Supplied by fuel | —— B.t.u. |
| Absorbed by water | —— B.t.u. |
| In exhaust gases | —— B.t.u. |
| Absorbed in work | —— B.t.u. |
| Radiation and friction | —— B.t.u. |
| Thermal efficiency | Per cent —— |
| B.t.u. per i.hp. | —— |

As there will probably be some items in the report not readily understood by the reader, an explanation of these will be given.

The ratio of gas to air is the quantity of gas consumed per hour divided by the air consumed. In the engines which take a charge of air into the cylinder, whether gas is taken or not, as in the type governed by means of the "hit or miss" cut-off type of governor, the exact ratio cannot be obtained, except as the engine takes an impulse at every cycle. An approximate result may be obtained, however, by taking the difference between the actual average cycles performed per hour and the average number of explosions per hour and subtracting from the air meter reading the computed amount of air used in these idle cycles, which would be the product of the cylinder volume times the idle cycles, and using this corrected value of air in the ratio. This ratio would not be absolutely accurate, owing to the probability of the cylinder not always obtaining a complete fresh charge of air on the suction stroke. The weight of the gas fuel may be obtained from the records of the gas company or it may be computed from the quantitative analysis of the gas — this analysis should be performed by a competent chemist. When the constituent parts of the gas, together with their weight, are known the weight may be determined by multiplying the weight per cubic foot of the constituents by their percentages as they appear in the gas analysis, and adding the results. Table XIV will give the weights and specific heats of the constituent parts of the gases most commonly encountered as gas-engine fuels.

TABLE XIV.

WEIGHT AND SPECIFIC HEAT OF GASES.

| Constituent. | Lb. per Cu. Ft. | Sp. Ht. Const. Pr. | Sp. Ht. Const. Vol. |
|---|---|---|---|
| Air | 0.08082 | 0.237 | 0.168 |
| Hydrogen | 0.00559 | 3.409 | 2.406 |
| Oxygen | 0.0894 | 0.217 | 0.155 |
| Nitrogen | 0.0779 | 0.244 | 0.173 |
| Marsh gas, $CH_4$ | 0.0445 | 0.593 | 0.467 |
| Carbonic oxide, CO | ........... | 0.245 | 0.173 |
| Carbonic acid, $CO_2$ | ........... | 0.216 | 0.171 |
| Olefines | ........... | 0.404 | 0.332 |

The weights and specific heats given are for an atmospheric pressure of 14.7 lb. per sq. in. and a temperature of 32 deg. fahr. The specific heat of the gas and of the mixture may be found by the same method as for finding the weight, the percentage of the constituent parts being known and their specific heat is found in the above table. The weight of the mixture is found by taking the weight of air to be 0.08082 lb. at 32 deg. fahr. and atmosphere 14.7. The weight of the gas being known, or having been determined, the weight of mixture per cubic foot is found by the following formula:

$$W_m = (0.08082)\, x + ay \quad . \quad . \quad . \quad (46)$$

In which $x$ is the percentage of air, $a$ the weight of the gas per cubic foot, and $y$ the percentage of gas.

The heating value of the gas should always be determined at a laboratory by an expert chemist. While Table I, Chapter IX, gives the heating values of various gases, their composition in different localities or under different conditions is subject to variation and for accurate results should not be depended on, but only used as a basis for computing probable results. Samples of the gas should be obtained at different times and then mixed; a sample of the mixture should then be sent to the laboratory.

The volumes of gas and air as obtained in the test and recorded in the log sheet must be reduced to standard temperature and atmospheric pressure in order to form a basis of comparison. The standards in use are the temperature of water at the freezing point, 32 deg. fahr., and the atmospheric pressure at sea level, which is equivalent to 30 in. of mercury. The following formula may be used in the reduction:

$$V = \frac{vp \times 491.2}{14.7 \times (t_1 + 459.2)} \quad . \quad . \quad . \quad (47)$$

In which

$t_1$ = temperature at time of test.
$p$ = atmospheric pressure at time of test.
$v$ = volume at this pressure and temperature.
$V$ = corrected volume at 32 deg. fahr. and pressure of 14.7 lb.

This formula, while derived on the basis of the air thermometer, will give an approximately correct value when used for the gas. The gas pressure, as measured in the test, in inches of water may be reduced to inches of mercury by dividing by 13.62, or if the pressure of the gas has been taken in pounds per sq. in. it may be reduced to inches of mercury by multiplying by 2.033. These ratios are for temperatures of 32 deg. fahr. but will be found to give sufficiently accurate results if used at any average temperature.

The indicated work is to be computed from the indicator cards taken during the test and is the product of the mean effective pressure, the area of the piston in inches, the stroke in feet, and the number of explosions per minute. The mean ordinate of the cards is obtained best by means of a planimeter.* If a planimeter is not available it may be obtained by the ordinate method, see Fig. 106.

From the atmospheric line *AB* erect ordinates equal distances apart, the first and last ordinates being half a space

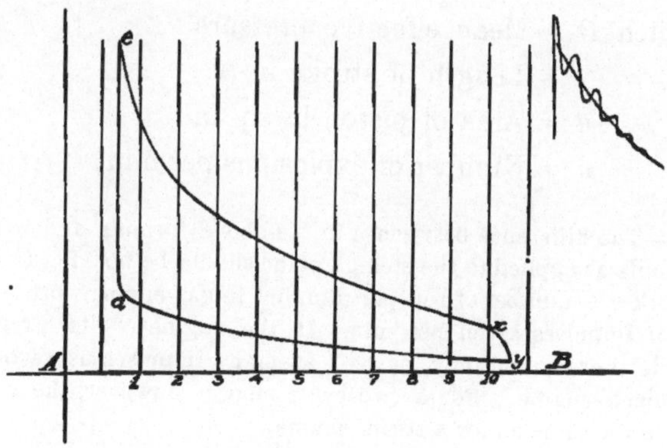


**Fig. 106. Engine Card.**

from the ends of the diagram. Add the lengths of the lines contained in the diagram, in this case ten, from *AB* to the expansion curve *ex*, and add the lengths of the same lines from *AB* to the compression curve *ay*, subtract the second sum from the first and divide by the number of lines. The result will be the approximate value of the

* A planimeter, as the name implies, is an instrument for computing areas. It may be so adjusted as to record, without further calculation, the mean ordinate of any irregular figure of given length. See Carpenter's "Experimental Engineering," page 31.

mean ordinate in inches. If the expansion curve should be irregular, due to the vibration of the indicator spring, a mean curve may be drawn as shown to right of Fig. 106. While the ordinate method will give fairly close results, its use in computations requiring extreme accuracy is not advisable.

The mean effective pressure is the result obtained by multiplying together the mean ordinate in inches and the scale of the spring in lb. per sq. in., which is known. The mean effective pressure being known, the indicated horsepower is calculated by means of the following formula:

$$H_i = \frac{Plan}{33,000} \quad . \quad . \quad . \quad . \quad . \quad . \quad (48)$$

In which $P$ = Mean effective pressure.
$l$ = Length of stroke in ft.
$a$ = Area of piston in sq. in.
$n$ = Number of explosions per min.

NOTE.—The difference between $n$ in the above formula and $n$ in the same formula as applied to the steam engine should be noted. In steam-engine work $n$ = number of rev. per min. but in gas-engine work $n$ = the number of impulses given per min. If the engine "hits" regularly every cycle there would only be half as many impulses as revolutions in a four-cycle engine. For a two-cycle engine, however, the value of $n$ would be the same as for a steam engine.

The quantity of heat supplied by the gas per hour is the product of the heating value per cu. ft. times the cu. ft. consumed. The heat absorbed in the water is the product of the water consumed, as determined in the test, and the range of temperature:

$$H_j = (t_2 - t_1) W \quad . \quad . \quad . \quad . \quad . \quad . \quad (49)$$

In which $t_1$ = The temperature of the discharge.
$t_2$ = Temperature of the inlet.
$W$ = Weight of the cooling water.

The result obtained is in B.t.u (British thermal units). The B.t.u. is the heat required to raise 1 lb. of water through 1 deg. fahr.

The heat absorbed in work is the ft.-lb. per hr. divided by the foot-pound equivalent of the B.t.u., which is 778:

$$H_w = \frac{2\,\pi r W n \times 60}{778} \quad . \quad . \quad . \quad . \quad . \quad (50)$$

The heat carried off in the exhaust is found as follows: the specific heat of the mixture having been determined, as previously described, gives us, in terms of a decimal, the B.t.u. required to raise one pound through one degree fahrenheit. Then knowing the weight of a cu. ft. of mixture in pounds, the quantity exhausted in cu. ft. and the range of temperature (obtained by means of the pyrometer and thermometer), the formula resolves itself into one of the same class as that derived for water:

$$H_e = S\,(t_2 - t_1)\,W \quad . \quad . \quad . \quad . \quad . \quad . \quad (51)$$

In which

$S$ = Specific heat of mixture.
$t_2$ = Temperature as obtained by pyrometer.
$t_1$ = Temperature of entering mixture or gas.
$W$ = Total weight of mixture per hour.

To be absolutely accurate the amount of unburned mixture passing out with the exhaust should be considered, and its heating value subtracted from that of the fuel entering the engine, but in practice it is considered as part of the heat lost by radiation.

The heat lost by radiation is determined by subtracting the sum of the three computed losses from the heat supplied by the gas.

The thermal efficiency of the engine is the quotient of the heat absorbed in work divided by the heat supplied the engine in fuel:

$$T = \frac{(2 r W n)\,60}{778\,CH} \quad . \quad . \quad . \quad . \quad . \quad . \quad . \quad (52)$$

In which the numerator is recognized as the numerator of the formula (50)

$C$ = Cu. ft. of gas per hr.
$H$ = Heat value per cu. ft.

The b. hp. of the engine is found by means of the formula derived for the prony brake, $H = \frac{2\pi rWn}{33,000}$.

In which the constants for different lengths of $r$ are given in the table, page 200.

## MISCELLANEOUS.

*The Muffler.* — The volume of muffler necessary, for any given size of engine, can be only approximately determined by formulas. As a general proposition, however, the volume of the muffler is from four to six times the total cylinder volume. While it is true that some engines, especially those in use on motorcycles, employ a muffler with a much less comparative volume, their use does not produce as good results from the standpoint of quietness of operation although the back pressure is less.

A muffler, in its simplest form, consists of an iron box or drum into which the exhaust gases are discharged before passing to the atmosphere. While such a muffler deadens, to a certain extent, the noise of the explosion, the development of the automobile industry has demanded a more efficient sound deadener. To produce this result the gases are made to pass through orifices or past baffle plates in order to give them more time to approach the pressure of the atmosphere before discharging into it. One simple method is to enclose an iron pipe, bored full of small holes, in an iron shell and discharge the exhaust gases into the pipe from which they pass through the holes to the shell of the muffler and finally to the open air.

It should be remembered that the more efficient a muffler is the greater is the back pressure and loss of power; hence the use of the simple drum type on industrial engines and other engines where quietness of operation is a secondary consideration.

## DEFINITIONS OF UNITS.

*Work.* — The sustained exertion of pressure through space.

*Unit of Work.* — One foot pound, i.e., a pressure of one pound exerted through a space of one foot.

*Horse-power.* — The rate of work. Unit of horse-power = 33,000 ft. lb. per minute, or 550 ft. lb. per second = 1,980,000 ft. lb. per hour.

*Heat Unit.* — Heat required to raise 1 lb. of water 1 deg. Fahr. (from 39 deg. to 40 deg.).

Horse-power expressed in heat units $= \frac{33,000}{778} = 42.416$ heat units per minute = 0.707 heat units per second = 2545 heat units per hour.

$$1 \text{ lb. of fuel per hp.-hr.} = \begin{cases} 1,980,000 \text{ ft. lb. per lb. of fuel.} \\ 2545 \text{ heat units.} \end{cases}$$

1,000,000 ft. lb. per lb. of fuel = 1.98 lb. of fuel per hp.-hr.

*Velocity.* — Feet per second $\frac{5280}{3600} = \frac{22}{15} \times$ miles per hour.

## TABLE XV.

### WIRE AND SHEET-METAL GAUGES COMPARED.

| Number of Gauge. | Birmingham or Stubs' Iron Wire. (Plate Metal.) | American or Browne & Sharpe. (Sheet Brass, Copper Tube.) | Roebling's and Washburn & Moen's. | Stubs' Steel Wire. | U. S. Std. Sheet & Plate Iron and Steel. | British Std. Wire Gauge. | | Piano Wire. |
|---|---|---|---|---|---|---|---|---|
| | inch | inch | inch | inch | inch | inch | mm. | inch |
| 0000000 | ...... | ...... | .49 | ...... | .5 | .500 | 12.7 | ...... |
| 000000 | ...... | ...... | .46 | ...... | .469 | .464 | 11.78 | ...... |
| 00000 | ...... | ...... | .43 | ...... | .438 | .432 | 10.97 | ...... |
| 0000 | .454 | .46 | .393 | ...... | .406 | .4 | 10.16 | ...... |
| 000 | .425 | .40964 | .362 | ...... | .375 | .372 | 9.45 | ...... |
| 00 | .38 | .3648 | .331 | ...... | .344 | .348 | 8.84 | .0085 |
| 0 | .34 | .32486 | .307 | ...... | .313 | .324 | 8.23 | .009 |
| 1 | .3 | .2893 | .283 | .227 | .281 | .3 | 7.62 | .010 |
| 2 | .284 | .25763 | .263 | .219 | .266 | .276 | 7.01 | .011 |
| 3 | .259 | .22942 | .244 | .212 | .25 | .252 | 6.4 | .012 |
| 4 | .238 | .20431 | .225 | .207 | .234 | .232 | 5.89 | .013 |
| 5 | .22 | .18194 | .207 | .204 | .219 | .212 | 5.38 | .014 |
| 6 | .203 | .16202 | .192 | .201 | .203 | .192 | 4.88 | .016 |
| 7 | .18 | .14428 | .177 | .199 | .188 | .176 | 4.47 | .018 |
| 8 | .165 | .12849 | .162 | .197 | .172 | .16 | 4.06 | .020 |
| 9 | .148 | .11443 | .148 | .194 | .156 | .144 | 3.66 | .022 |
| 10 | .134 | .10189 | .135 | .191 | .141 | .128 | 3.25 | .024 |

WIRE AND SHEET-METAL GAUGES COMPARED. — *Continued.*

| Number of Gauge. | Birmingham or Stubs' Iron Wire. (Plate Metal.) | American or Browne & Sharpe. (Sheet Brass, Copper Tube.) | Roebling's and Washburn & Moen's. | Stubs' Steel Wire. | U. S. Std. Sheet and Plate Iron and Steel. | British Std. Wire Gauge. | | Piano Wire. |
|---|---|---|---|---|---|---|---|---|
| | inch | inch | inch | inch | inch | inch | inch | inch |
| 11 | .12 | .09074 | .120 | .188 | .125 | .116 | 2.95 | .026 |
| 12 | .109 | .08081 | .105 | .185 | .109 | .104 | 2.64 | .028 |
| 13 | .095 | .07196 | .092 | .182 | .094 | .092 | 2.34 | .030 |
| 14 | .083 | .06408 | .080 | .180 | .078 | .08 | 2.03 | .032 |
| 15 | .072 | .05707 | .072 | .178 | .07 | .072 | 1.83 | .034 |
| 16 | .065 | .05082 | .063 | .175 | .0625 | .064 | 1.63 | .036 |
| 17 | .058 | .04526 | .054 | .172 | .0563 | .056 | 1.42 | .038 |
| 18 | .049 | .0403 | .047 | .168 | .05 | .048 | 1.22 | .040 |
| 19 | .042 | .03589 | .041 | .164 | .0438 | .04 | 1.02 | .042 |
| 20 | .035 | .03196 | .035 | .161 | .0375 | .036 | .91 | .044 |
| 21 | .032 | .02846 | .032 | .157 | .0344 | .032 | .81 | .046 |
| 22 | .028 | .02535 | .028 | .155 | .0313 | .028 | .71 | .048 |
| 23 | .025 | .02257 | .025 | .153 | .0281 | .024 | .61 | .051 |
| 24 | .022 | .0201 | .023 | .151 | .025 | .022 | .56 | .055 |
| 25 | .02 | .0179 | .020 | .148 | .0219 | .02 | .51 | .059 |
| 26 | .018 | .01594 | .018 | .146 | .0188 | .018 | .46 | .063 |
| 27 | .016 | .01419 | .017 | .143 | .0172 | .0164 | .42 | .067 |
| 28 | .014 | .01264 | .016 | .139 | .0156 | .0148 | .38 | .071 |
| 29 | .013 | .01126 | .015 | .134 | .0141 | .0136 | .35 | .074 |
| 30 | .012 | .01002 | .014 | .127 | .0125 | .0124 | .31 | .078 |
| 31 | .01 | .00893 | .0135 | .120 | .0109 | .0116 | .29 | .082 |
| 32 | .009 | .00795 | .013 | .115 | .0101 | .0108 | .27 | .086 |
| 33 | .008 | .00708 | .011 | .112 | .0094 | .01 | .25 | .090 |
| 34 | .007 | .0063 | .01 | .110 | .0086 | .0092 | .23 | .094 |
| 35 | .005 | .00561 | .0095 | .108 | .0078 | .0084 | .21 | .098 |
| 36 | .004 | .005 | .009 | .106 | .007 | .0076 | .19 | ...... |
| 37 | ...... | .00445 | .0085 | .103 | .0066 | .0068 | .17 | ...... |
| 38 | ...... | .00396 | .008 | .101 | .0063 | .006 | .15 | ...... |
| 39 | ...... | .00353 | .0075 | .099 | ...... | .0052 | .13 | ...... |
| 40 | ...... | .00314 | .007 | .097 | ...... | .0048 | .12 | ...... |
| 41 | ...... | ...... | ...... | .095 | ...... | .0044 | .11 | ...... |
| 42 | ...... | ...... | ...... | .092 | ...... | .004 | .10 | ...... |
| 43 | ...... | ...... | ...... | .088 | ...... | .0036 | .09 | ...... |
| 44 | ...... | ...... | ...... | .085 | ...... | .0032 | .08 | ...... |
| 45 | ...... | ...... | ...... | .081 | ...... | .0028 | .07 | ...... |
| 46 | ...... | ...... | ...... | .079 | ...... | .0024 | .06 | ...... |
| 47 | ...... | ...... | ...... | .077 | ...... | .002 | .05 | ...... |
| 48 | ...... | ...... | ...... | .075 | ...... | .0016 | .04 | ...... |
| 49 | ...... | ...... | ...... | .072 | ...... | .0012 | .03 | ...... |
| 50 | ...... | ...... | ...... | .069 | ...... | .001 | .025 | ...... |

## TABLE XVI.

### TAP DRILL TABLE.

| Diameter of Taps. | Number of Threads per Inch. | | Diameter of Tap Drill. | |
|---|---|---|---|---|
| | V. Thread. | U. S. Std. Thread. | V. Thread. | U. S. Std. Thread. |
| 1/8 | 32 | ........ | #44 | ........... |
| 3/16 | 24 | ........ | #29 | ........... |
| 1/4 | 20 | 20 | #14 | 3/16 |
| 5/16 | 18 | 18 | 15/64 | C |
| 3/8 | 16 | 18 | L 9/32 | N |
| 7/16 | 14 | 14 | R | S |
| 1/2 | 12 | 13 | 25/64 | 13/32 |
| 9/16 | 12 | 12 | 29/64 | 29/64 |
| 5/8 | 11 | 11 | 1/2 | 33/64 |
| 11/16 | 11 | 11 | 9/16 | 37/64 |
| 3/4 | 10 | 10 | 39/64 | 5/8 |
| 13/16 | 10 | 10 | 48/64 | 11/16 |
| 7/8 | 9 | 9 | 23/32 | 47/64 |
| 15/16 | 9 | 9 | 25/32 | 51/64 |
| 1 | 8 | 8 | 53/64 | 27/32 |
| 1/8 | 7 | 7 | 59/64 | 61/64 |
| 1/4 | 7 | 7 | 1 3/64 | 1 5/64 |
| 3/8 | 6 | 6 | 1 1/8 | 1 11/64 |
| 1/2 | 6 | 6 | 1 17/64 | 1 19/64 |
| 5/8 | 5 | 5 1/2 | 1 21/64 | 1 25/64 |
| 3/4 | 5 | 5 | 1 29/64 | 1 1/2 |
| 7/8 | 4 1/2 | 5 | 1 35/64 | 1 5/8 |
| 2 | 4 1/2 | 4 1/2 | 1 43/64 | 1 23/32 |

# TABLE XVII.

## MACHINE SCREW TABLE.

| Screw Gauge. | Diam. Decimals. | Approx. Diam. | No. Thread per 1 Inch. | Size Tap Dr. |
|---|---|---|---|---|
| 2 | .0842 | $\frac{5}{64}$ | 56 | 49 |
| 3 | .0973 | $\frac{3}{32}$ | 48 | 45 |
| 4 | .1105 | $\frac{7}{64}$ | 36 | 42 |
| 5 | .1236 | $\frac{1}{8}$ | 36 | 38 |
| 6 | .1368 | $\frac{9}{64}$ | 32 | 35 |
| 7 | .1500 | $\frac{5}{32}$ | 32 | 30 |
| 8 | .1631 | $\frac{5}{32}$ | 32 | 29 |
| 9 | .1763 | $\frac{11}{64}$ | 30 | 27 |
| 10 | .1894 | $\frac{3}{16}$ | 24 | 25 |
| 11 | .2026 | $\frac{13}{64}$ | 24 | 21 |
| 12 | .2156 | $\frac{7}{32}$ | 24 | 17 |
| 13 | .2289 | $\frac{15}{64}$ | 22 | 15 |
| 14 | .2421 | $\frac{15}{64}$ | 20 | 13 |
| 15 | .2552 | $\frac{1}{4}$ | 20 | 8 |
| 16 | .2634 | $\frac{17}{64}$ | 18 | 5 |
| 17 | .2816 | $\frac{9}{32}$ | 18 | 2 |
| 18 | .2947 | $\frac{19}{64}$ | 18 | 1 |
| 19 | .3079 | $\frac{5}{16}$ | 18 | C |
| 20 | .3210 | $\frac{21}{64}$ | 16 | D |
| 22 | .3474 | $\frac{11}{32}$ | 16 | J |
| 24 | .3737 | $\frac{3}{8}$ | 16 | N |
| 26 | .4000 | $\frac{13}{32}$ | 16 | P |
| 28 | .4263 | $\frac{27}{64}$ | 14 | R |
| 30 | .4526 | $\frac{29}{64}$ | 14 | U |

## TABLE XVIII.

### WROUGHT IRON PIPE.

| Diameter. | | | Thread per Inch. | Size of Tap Drill. |
|---|---|---|---|---|
| Nominal. | Actual External. | Actual Internal. | | |
| 1/8 | .405 | .27 | 27 | 21/64 |
| 1/4 | .540 | .364 | 18 | 29/64 |
| 3/8 | .675 | .494 | 18 | 19/32 |
| 1/2 | .840 | .623 | 14 | 23/32 |
| 3/4 | 1.050 | .824 | 14 | 15/16 |
| 1 | 1.315 | 1.048 | 11 1/2 | 1 3/16 |
| 1/4 | 1.660 | 1.38 | 11 1/2 | 1 15/32 |
| 1/2 | 1.900 | 1.611 | 11 1/2 | 1 23/32 |
| 2 | 2.375 | 2.067 | 11 1/2 | 2 3/16 |
| 1/2 | 2.875 | 2.468 | 8 | 2 5/8 |
| 3 | 3.500 | 3.067 | 8 | 3 1/4 |
| 1/2 | 4.0 | 3.548 | 8 | 3 3/4 |
| 4 | 4.500 | 4.026 | 8 | 4 1/4 |
| 1/2 | 5.0 | 4.508 | 8 | 4 3/4 |
| 5 | 5.563 | 5.045 | 8 | 5 5/16 |
| 6 | 6.625 | 6.065 | 8 | 6 5/16 |
| 7 | 7.625 | 7.023 | 8 | 7 3/8 |
| 8 | 8.625 | 7.982 | 8 | 8 3/8 |
| 9 | 9.625 | 8.937 | 8 | .......... |
| 10 | 10.75 | 10.019 | 8 | .......... |

## TABLE XIX.—CIRCUMFERENCES AND AREAS OF CIRCLES ADVANCING BY EIGHTS.

| Diam. | Circ. | Area. | Diam. | Circ. | Area. | Diam. | Circ. | Area. |
|---|---|---|---|---|---|---|---|---|
| 0 1/64 | .04909 | .00019 | 2 | 6.2832 | 3.1416 | 5 | 15.708 | 19.635 |
| 1/32 | .09818 | .00077 | 1/16 | 6.4795 | 3.3410 | 1/16 | 15.904 | 20.129 |
| 3/64 | .14726 | .00173 | 1/8 | 6.6759 | 3.5466 | 1/8 | 16.101 | 20.629 |
| 1/16 | .19635 | .00307 | 3/16 | 6.8722 | 3.7583 | 3/16 | 16.297 | 21.135 |
| 3/32 | .29452 | .00690 | 1/4 | 7.0686 | 3.9761 | 1/4 | 16.493 | 21.648 |
| 1/8 | .39270 | .01227 | 5/16 | 7.2649 | 4.2000 | 5/16 | 16.690 | 22.166 |
| 5/32 | .49087 | .01917 | 3/8 | 7.4613 | 4.4301 | 3/8 | 16.886 | 22.691 |
| 3/16 | .58905 | .02761 | 7/16 | 7.6576 | 4.6664 | 7/16 | 17.082 | 23.221 |
| 7/32 | .68722 | .03758 | | | | | | |
| 1/4 | .78540 | .04909 | 1/2 | 7.8540 | 4.9087 | 1/2 | 17.279 | 23.758 |
| 9/32 | .88357 | .06213 | 9/16 | 8.0503 | 5.1572 | 9/16 | 17.475 | 24.301 |
| 5/16 | .98175 | .07670 | 5/8 | 8.2467 | 5.4119 | 5/8 | 17.671 | 24.850 |
| 11/32 | 1.0799 | .09281 | 11/16 | 8.4430 | 5.6727 | 11/16 | 17.868 | 25.406 |
| 3/8 | 1.1781 | .11045 | 3/4 | 8.6394 | 5.9396 | 3/4 | 18.064 | 25.967 |
| 13/32 | 1.2763 | .12962 | 13/16 | 8.8357 | 6.2126 | 13/16 | 18.261 | 26.535 |
| 7/16 | 1.3744 | .15033 | 7/8 | 9.0321 | 6.4918 | 7/8 | 18.457 | 27.109 |
| 15/32 | 1.4726 | .17257 | 15/16 | 9.2284 | 6.7771 | 15/16 | 18.653 | 27.688 |
| 1/2 | 1.5708 | .19635 | 3 | 9.4248 | 7.0686 | 6 | 18.850 | 28.274 |
| 17/32 | 1.6690 | .22166 | 1/16 | 9.6211 | 7.3662 | 1/8 | 19.242 | 29.465 |
| 9/16 | 1.7671 | .24850 | 1/8 | 9.8175 | 7.6999 | 1/4 | 19.635 | 30.680 |
| 19/32 | 1.8653 | .27688 | 3/16 | 10.014 | 7.9798 | 3/8 | 20.028 | 31.919 |
| 5/8 | 1.9635 | .30680 | 1/4 | 10.210 | 8.2958 | 1/2 | 20.420 | 33.183 |
| 21/32 | 2.0617 | .33824 | 5/16 | 10.407 | 8.6179 | 5/8 | 20.813 | 34.472 |
| 11/16 | 2.1598 | .37122 | 3/8 | 10.603 | 8.9462 | 3/4 | 21.206 | 35.785 |
| 23/32 | 2.2580 | .40574 | 7/16 | 10.799 | 9.2806 | 7/8 | 21.598 | 37.122 |
| 3/4 | 2.3562 | .44179 | 1/2 | 10.996 | 9.6211 | 7 | 21.991 | 38.485 |
| 25/32 | 2.4544 | .47937 | 9/16 | 11.192 | 9.9678 | 1/8 | 22.384 | 39.871 |
| 13/16 | 2.5525 | .51849 | 5/8 | 11.388 | 10.321 | 1/4 | 22.776 | 41.282 |
| 27/32 | 2.6507 | .55914 | 11/16 | 11.585 | 10.680 | 3/8 | 23.169 | 42.718 |
| 7/8 | 2.7489 | .60132 | 3/4 | 11.781 | 11.045 | 1/2 | 23.562 | 44.179 |
| 29/32 | 2.8471 | .64504 | 13/16 | 11.977 | 11.416 | 5/8 | 23.955 | 45.664 |
| 15/16 | 2.9452 | .69029 | 7/8 | 12.174 | 11.793 | 3/4 | 24.347 | 47.173 |
| 31/32 | 3.0434 | .73708 | 15/16 | 12.370 | 12.177 | 7/8 | 24.740 | 48.707 |
| 1 | 3.1416 | .7854 | 4 | 12.566 | 12 566 | 8 | 25.133 | 50.265 |
| 1/16 | 3.3379 | .8866 | 1/16 | 12.763 | 12.962 | 1/8 | 25.525 | 51.849 |
| 1/8 | 3.5343 | .9940 | 1/8 | 12.959 | 13.364 | 1/4 | 25.918 | 53.456 |
| 3/16 | 3.7306 | 1.1075 | 3/16 | 13.155 | 13.772 | 3/8 | 26.311 | 55.088 |
| 1/4 | 3.9270 | 1.2272 | 1/4 | 13.352 | 14.186 | 1/2 | 26.704 | 56.745 |
| 5/16 | 4.1233 | 1.3530 | 5/16 | 13.548 | 14.607 | 5/8 | 27.096 | 58.426 |
| 3/8 | 4.3197 | 1.4849 | 3/8 | 13.744 | 15.033 | 3/4 | 27.489 | 60.132 |
| 7/16 | 4.5160 | 1.6230 | 7/16 | 13.941 | 15.466 | 7/8 | 27.882 | 61.862 |
| 1/2 | 4.7124 | 1.7671 | 1/2 | 14.137 | 15.904 | 9 | 28.274 | 63.617 |
| 9/16 | 4.9087 | 1.9175 | 9/16 | 14.334 | 16.349 | 1/8 | 28.667 | 65.397 |
| 5/8 | 5.1051 | 2.0739 | 5/8 | 14.530 | 16.800 | 1/4 | 29.060 | 67.201 |
| 11/16 | 5.3014 | 2.2365 | 11/16 | 14.726 | 17.257 | 3/8 | 29.452 | 69.029 |
| 3/4 | 5.4978 | 2.4053 | 3/4 | 14.923 | 17.728 | 1/2 | 29.845 | 70.882 |
| 13/16 | 5.6941 | 2.5802 | 13/16 | 15.119 | 18.190 | 5/8 | 30.238 | 72.760 |
| 7/8 | 5.8905 | 2.7612 | 7/8 | 15.315 | 18.665 | 3/4 | 30.631 | 74.662 |
| 15/16 | 6.0868 | 2.9483 | 15/16 | 15.512 | 19.147 | 7/8 | 31.023 | 76.589 |

CIRCUMFERENCES AND AREAS OF CIRCLES.—*Continued.*

| Diam. | Circ. | Area. | Diam. | Circ. | Area. | Diam. | Circ. | Area. |
|---|---|---|---|---|---|---|---|---|
| **10** | 31.416 | 78.540 | **16** | 50.265 | 201. | **22** | 69.115 | 380.13 |
| 1/8 | 31.809 | 80.516 | 1/8 | 50.658 | 204. | 1/8 | 69.508 | 384.46 |
| 1/4 | 32.201 | 82.516 | 1/4 | 51.051 | 207. | 1/4 | 69.900 | 388.82 |
| 3/8 | 32.594 | 84.541 | 3/8 | 51.444 | 210. | 3/8 | 70.293 | 393.20 |
| 1/2 | 32.987 | 86.590 | 1/2 | 51.836 | 213. | 1/2 | 70.686 | 397.61 |
| 5/8 | 33.379 | 88.664 | 5/8 | 52.229 | 217. | 5/8 | 71.079 | 402.04 |
| 3/4 | 33.772 | 90.763 | 3/4 | 52.622 | 220. | 3/4 | 71.471 | 406.49 |
| 7/8 | 34.165 | 92.886 | 7/8 | 53.014 | 223. | 7/8 | 71.864 | 410.97 |
| **11** | 34.558 | 95.033 | **17** | 53.407 | 226.98 | **23** | 72.257 | 415.48 |
| 1/8 | 34.950 | 97.205 | 1/8 | 53.800 | 230.33 | 1/8 | 72.649 | 420.00 |
| 1/4 | 35.343 | 99.402 | 1/4 | 54.192 | 233.71 | 1/4 | 73.042 | 424.56 |
| 3/8 | 35.736 | 101.62 | 3/8 | 54.585 | 237.10 | 3/8 | 73.435 | 429.13 |
| 1/2 | 36.128 | 103.87 | 1/2 | 54.978 | 240.53 | 1/2 | 73.827 | 433.74 |
| 5/8 | 36.521 | 106.14 | 5/8 | 55.371 | 243.98 | 5/8 | 74.220 | 438.36 |
| 3/4 | 36.914 | 108.43 | 3/4 | 55.763 | 247.45 | 3/4 | 74.613 | 443.01 |
| 7/8 | 37.306 | 110.75 | 7/8 | 56.156 | 250.95 | 7/8 | 75.006 | 447.69 |
| **12** | 37.699 | 113.10 | **18** | 56.549 | 254.47 | **24** | 75.398 | 452.39 |
| 1/8 | 38.092 | 115.47 | 1/8 | 56.941 | 258.02 | 1/8 | 75.791 | 457.11 |
| 1/4 | 38.485 | 117.86 | 1/4 | 57.334 | 261.59 | 1/4 | 76.184 | 461.86 |
| 3/8 | 38.877 | 120.28 | 3/8 | 57.727 | 265.18 | 3/8 | 76.576 | 466.64 |
| 1/2 | 39.270 | 122.72 | 1/2 | 58.119 | 268.80 | 1/2 | 76.696 | 471.44 |
| 5/8 | 39.663 | 125.19 | 5/8 | 58.512 | 272.45 | 5/8 | 77.362 | 476.26 |
| 3/4 | 40.055 | 127.68 | 3/4 | 58.905 | 276.12 | 3/4 | 77.754 | 481.11 |
| 7/8 | 40.448 | 130.19 | 7/8 | 59.298 | 279.81 | 7/8 | 78.147 | 485.98 |
| **13** | 40.841 | 132.73 | **19** | 59.690 | 283.53 | **25** | 78.540 | 490.87 |
| 1/8 | 41.233 | 135.30 | 1/8 | 60.083 | 287.27 | 1/8 | 78.933 | 495.79 |
| 1/4 | 41.626 | 137.89 | 1/4 | 60.476 | 291.04 | 1/4 | 79.325 | 500.74 |
| 3/8 | 42.019 | 140.50 | 3/8 | 60.868 | 294.83 | 3/8 | 79.718 | 505.71 |
| 1/2 | 42.412 | 143.14 | 1/2 | 61.261 | 298.65 | 1/2 | 80.111 | 510.71 |
| 5/8 | 42.804 | 145.80 | 5/8 | 61.654 | 302.49 | 5/8 | 80.503 | 515.72 |
| 3/4 | 43.197 | 148.49 | 3/4 | 62.046 | 306.35 | 3/4 | 80.896 | 520.77 |
| 7/8 | 43.590 | 151.20 | 7/8 | 62.439 | 310.24 | 7/8 | 81.289 | 525.84 |
| **14** | 43.982 | 153.94 | **20** | 62.832 | 314.16 | **26** | 81.681 | 530.93 |
| 1/8 | 44.375 | 156.70 | 1/8 | 63.225 | 318.10 | 1/8 | 82.074 | 536.05 |
| 1/4 | 44.768 | 159.48 | 1/4 | 63.617 | 322.06 | 1/4 | 82.467 | 541.19 |
| 3/8 | 45.160 | 162.30 | 3/8 | 64.010 | 326.05 | 3/8 | 82.860 | 546.35 |
| 1/2 | 45.553 | 165.13 | 1/2 | 64.403 | 330.06 | 1/2 | 83.252 | 551.55 |
| 5/8 | 45.946 | 167.99 | 5/8 | 64.795 | 334.10 | 5/8 | 83.645 | 556.76 |
| 3/4 | 46.338 | 170.87 | 3/4 | 65.188 | 338.16 | 3/4 | 84.038 | 562.00 |
| 7/8 | 46.731 | 173.78 | 7/8 | 65.581 | 342.25 | 7/8 | 84.430 | 567.27 |
| **15** | 47.124 | 176.71 | **21** | 65.973 | 346.36 | **27** | 84.823 | 572.56 |
| 1/8 | 47.517 | 179.67 | 1/8 | 66.366 | 350.50 | 1/8 | 85.216 | 577.87 |
| 1/4 | 47.909 | 182.65 | 1/4 | 66.759 | 354.66 | 1/4 | 85.608 | 583.21 |
| 3/8 | 48.302 | 185.66 | 3/8 | 67.152 | 358.84 | 3/8 | 86.001 | 588.57 |
| 1/2 | 48.695 | 188.69 | 1/2 | 67.544 | 363.05 | 1/2 | 86.394 | 593.96 |
| 5/8 | 49.087 | 191.75 | 5/8 | 67.937 | 367.28 | 5/8 | 86.786 | 599.37 |
| 3/4 | 49.480 | 194.83 | 3/4 | 68.330 | 371.54 | 3/4 | 87.179 | 604.81 |
| 7/8 | 49.873 | 197.93 | 7/8 | 68.722 | 375.83 | 7/8 | 87 572 | 610.27 |

## CIRCUMFERENCES AND AREAS OF CIRCLES. — *Continued.*

| Diam. | Circ. | Area. | Diam. | Circ. | Area. | Diam. | Circ. | Area. |
|---|---|---|---|---|---|---|---|---|
| **28** | 87.965 | 615.75 | **34** | 106.814 | 907.92 | **40** | 125.664 | 1256.6 |
| 1/8 | 88.357 | 621.26 | 1/8 | 107.207 | 914.61 | 1/8 | 126.056 | 1264.5 |
| 1/4 | 88.750 | 626.80 | 1/4 | 107.600 | 921.32 | 1/4 | 126.449 | 1272.4 |
| 3/8 | 89.143 | 632.36 | 3/8 | 107.992 | 928.06 | 3/8 | 126.842 | 1280.3 |
| 1/2 | 89.535 | 637.94 | 1/2 | 108.385 | 934.82 | 1/2 | 127.235 | 1288.2 |
| 5/8 | 89.928 | 643.55 | 5/8 | 108.778 | 941.61 | 5/8 | 127.627 | 1296.2 |
| 3/4 | 90.321 | 649.18 | 3/4 | 109.170 | 948.42 | 3/4 | 128.020 | 1304.2 |
| 7/8 | 90.713 | 654.84 | 7/8 | 109.563 | 955.25 | 7/8 | 128.413 | 1312.2 |
| **29** | 91.106 | 660.52 | **35** | 109.956 | 962.11 | **41** | 128.805 | 1320.3 |
| 1/8 | 91.499 | 666.23 | 1/8 | 110.348 | 969.00 | 1/8 | 129.198 | 1328.3 |
| 1/4 | 91.892 | 671.96 | 1/4 | 110.741 | 975.91 | 1/4 | 129.591 | 1336.4 |
| 3/8 | 92.284 | 677.71 | 3/8 | 111.134 | 982.84 | 3/8 | 129.983 | 1344.5 |
| 1/2 | 92.677 | 683.49 | 1/2 | 111.527 | 989.80 | 1/2 | 130.376 | 1352.7 |
| 5/8 | 93.070 | 689.30 | 5/8 | 111.919 | 996.78 | 5/8 | 130.769 | 1360.8 |
| 3/4 | 93.462 | 695.13 | 3/4 | 112.312 | 1003.8 | 3/4 | 131.161 | 1369.0 |
| 7/8 | 93.855 | 700.98 | 7/8 | 112.705 | 1010.8 | 7/8 | 131.554 | 1377.2 |
| **30** | 94.248 | 706.86 | **36** | 113.097 | 1017.9 | **42** | 131.947 | 1385.4 |
| 1/8 | 94.640 | 712.76 | 1/8 | 113.490 | 1025.0 | 1/8 | 132.340 | 1393.7 |
| 1/4 | 95.033 | 718.69 | 1/4 | 113.883 | 1032.1 | 1/4 | 132.732 | 1402.0 |
| 3/8 | 95.426 | 724.64 | 3/8 | 114.275 | 1039.2 | 3/8 | 133.125 | 1410.3 |
| 1/2 | 95.819 | 730.62 | 1/2 | 114.668 | 1046.3 | 1/2 | 133.518 | 1418.6 |
| 5/8 | 96.211 | 736.62 | 5/8 | 115.061 | 1053.5 | 5/8 | 133.910 | 1427.0 |
| 3/4 | 96.604 | 742.64 | 3/4 | 115.454 | 1060.7 | 3/4 | 134.303 | 1435.4 |
| 7/8 | 96.997 | 748.69 | 7/8 | 115.846 | 1068.0 | 7/8 | 134.696 | 1443.8 |
| **31** | 97.389 | 754.77 | **37** | 116.239 | 1075.2 | **43** | 135.088 | 1452.2 |
| 1/8 | 97.782 | 760.87 | 1/8 | 116.632 | 1082.5 | 1/8 | 135.481 | 1460.7 |
| 1/4 | 98.175 | 766.99 | 1/4 | 117.024 | 1089.8 | 1/4 | 135.874 | 1469.1 |
| 3/8 | 98.567 | 773.14 | 3/8 | 117.417 | 1097.1 | 3/8 | 136.267 | 1477.6 |
| 1/2 | 98.960 | 779.31 | 1/2 | 117.810 | 1104.5 | 1/2 | 136.659 | 1486.2 |
| 5/8 | 99.353 | 785.51 | 5/8 | 118.202 | 1111.8 | 5/8 | 137.052 | 1494.7 |
| 3/4 | 99.746 | 791.73 | 3/4 | 118.596 | 1119.2 | 3/4 | 137.445 | 1503.3 |
| 7/8 | 100.138 | 797.98 | 7/8 | 118.988 | 1126.7 | 7/8 | 137.837 | 1511.9 |
| **32** | 100.531 | 804.25 | **38** | 119.381 | 1134.1 | **44** | 138.230 | 1520.5 |
| 1/8 | 100.924 | 810.54 | 1/8 | 119.773 | 1141.0 | 1/8 | 138.623 | 1529.2 |
| 1/4 | 101.316 | 816.86 | 1/4 | 120.166 | 1149.1 | 1/4 | 139.015 | 1537.9 |
| 3/8 | 101.709 | 823.21 | 3/8 | 120.559 | 1156.6 | 3/8 | 139.408 | 1546.6 |
| 1/2 | 102.102 | 829.58 | 1/2 | 120.951 | 1164.2 | 1/2 | 139.801 | 1555.3 |
| 5/8 | 102.494 | 835.97 | 5/8 | 121.344 | 1171.7 | 5/8 | 140.194 | 1564.0 |
| 3/4 | 102.887 | 842.39 | 3/4 | 121.737 | 1179.3 | 3/4 | 140.586 | 1572.8 |
| 7/8 | 103.280 | 848.83 | 7/8 | 122.129 | 1186.9 | 7/8 | 140.979 | 1581.6 |
| **33** | 103.673 | 855.30 | **39** | 122.522 | 1194.6 | **45** | 141.372 | 1590.4 |
| 1/8 | 104.065 | 861.79 | 1/8 | 122.915 | 1202.3 | 1/8 | 141.764 | 1599.3 |
| 1/4 | 104.458 | 868.31 | 1/4 | 123.308 | 1210.0 | 1/4 | 142.157 | 1608.2 |
| 3/8 | 104.851 | 874.85 | 3/8 | 123.700 | 1217.7 | 3/8 | 142.550 | 1617.0 |
| 1/2 | 105.243 | 881.41 | 1/2 | 124.093 | 1225.4 | 1/2 | 142.942 | 1626.0 |
| 5/8 | 105.636 | 888.00 | 5/8 | 124.486 | 1233.2 | 5/8 | 143.335 | 1634.9 |
| 3/4 | 106.029 | 894.62 | 3/4 | 124.878 | 1241.0 | 3/4 | 143.728 | 1643.9 |
| 7/8 | 106.421 | 901.26 | 7/8 | 125.271 | 1248.8 | 7/8 | 144.121 | 1652 9 |

CIRCUMFERENCES AND AREAS OF CIRCLES.—*Continued.*

| Diam. | Circ. | Area. | Diam. | Circ. | Area. | Diam. | Circ. | Area. |
|---|---|---|---|---|---|---|---|---|
| **46** | 144.513 | 1661.9 | $\frac{3}{8}$ | 148.833 | 1762.7 | $\frac{3}{4}$ | 153.153 | 1866.5 |
| $\frac{1}{8}$ | 144.906 | 1670.9 | $\frac{1}{2}$ | 149.226 | 1772.1 | $\frac{7}{8}$ | 153.545 | 1876.1 |
| $\frac{1}{4}$ | 145.299 | 1680.0 | $\frac{5}{8}$ | 149.618 | 1781.4 | | | |
| $\frac{3}{8}$ | 145.691 | 1689.1 | $\frac{3}{4}$ | 150.011 | 1790.8 | **49** | 153.938 | 1885.7 |
| $\frac{1}{2}$ | 146.084 | 1698.2 | $\frac{7}{8}$ | 150.404 | 1800.1 | $\frac{1}{8}$ | 154.331 | 1895.4 |
| $\frac{5}{8}$ | 146.477 | 1707.4 | | | | $\frac{1}{4}$ | 154.723 | 1905.0 |
| $\frac{3}{4}$ | 146.869 | 1716.5 | **48** | 150.796 | 1809.6 | $\frac{3}{8}$ | 155.116 | 1914.7 |
| $\frac{7}{8}$ | 147.262 | 1725.7 | $\frac{1}{8}$ | 151.189 | 1819.0 | $\frac{1}{2}$ | 155.509 | 1924.4 |
| | | | $\frac{1}{4}$ | 151.582 | 1828.5 | $\frac{5}{8}$ | 155.902 | 1934.2 |
| **47** | 147.655 | 1734.9 | $\frac{3}{8}$ | 151.975 | 1837.9 | $\frac{3}{4}$ | 156.294 | 1943.9 |
| $\frac{1}{8}$ | 148.048 | 1744.2 | $\frac{1}{2}$ | 152.367 | 1847.5 | $\frac{7}{8}$ | 156.687 | 1953.7 |
| $\frac{1}{4}$ | 148.440 | 1753 5 | $\frac{5}{8}$ | 152.760 | 1857 0 | | | |

# TABLE XX.

## TRIGONOMETRIC TABLES.

To find the function of any angle between the values given in the table the following example will illustrate the method:

To find sin – 81° – 43′

sin – 81° – 40′ = 0.9894

sin – 81° – 50′ = 0.9899

Dif. last unit = 5

From table of proportionate parts, marked P. P., we have

$5 \times .3 = 1.5$

Adding 0.9894

we have sin 81° – 43′ = 0.98955

Where the difference, in last unit place, for 10 minutes is less than 4 the proportionate parts have not been tabulated, but are readily determined.

For angles more than 90°

sin, cos, tan, cotan, sec, or cosec of $90° + x°$

= sin, cos, tan, cotan, sec, or cosec of $180° - (90° + x)$.

sin, cosin, tan, cotan, sec, or cosec of $180° + x°$ (but less than 270°)

= sin, cosin, tan, cotan, sec, or cosec of $x°$.

sin, cosin, tan, cotan, sec, or cosec of $270° + x°$

= sin, cosin, tan, cotan, sec, or cosec of $360° - (270° + x°)$.

The sec and cosec, versed sin and co-versed sin are not contained in the following table, but their values may be determined as follows:

$$\text{sec} = \frac{1}{\text{cosin}} \qquad \text{versed. sin} = 1 - \text{cosin}.$$

$$\text{cosec} = \frac{1}{\text{sine}} \qquad \text{co-versed sin} = 1 - \text{sin}.$$

## TABLE XX.—TRIGONOMETRIC FUNCTIONS.

| ° | ′ | Sin. | d. | Tan. | d. | Cot. | d. | Cos. | d. | ′ | ° |
|---|---|---|---|---|---|---|---|---|---|---|---|
| 0 | 0 | 0.0000 | | 0.0000 | | infinit. | | 1.0000 | | 0 | 90 |
| | 10 | 0.0029 | 29 | 0.0029 | 29 | 343.7737 | | 1.0000 | 0 | 50 | |
| | 20 | 0.0058 | 29 | 0.0058 | 29 | 171.8854 | | 1.0000 | 0 | 40 | |
| | 30 | 0.0087 | 29 | 0.0087 | 29 | 114.5887 | | 1.0000 | 0 | 30 | |
| | 40 | 0.0116 | 29 | 0.0116 | 29 | 85.9398 | | 0.9999 | 1 | 20 | |
| | 50 | 0.0145 | 29 | 0.0145 | 29 | 68.7501 | | 0.9999 | 0 | 10 | |
| 1 | 0 | 0.0175 | 30 | 0.0175 | 30 | 57.2900 | | 0.9998 | 1 | 0 | 89 |
| | 10 | 0.0204 | 29 | 0.0204 | 29 | 49.1039 | 81861 | 0.9998 | 0 | 50 | |
| | 20 | 0.0233 | 29 | 0.0233 | 29 | 42.9641 | 61398 | 0.9997 | 1 | 40 | |
| | 30 | 0.0262 | 29 | 0.0262 | 29 | 38.1885 | 47756 | 0.9997 | 0 | 30 | |
| | 40 | 0.0291 | 29 | 0.0291 | 29 | 34.3678 | 38207 | 0.9996 | 1 | 20 | |
| | 50 | 0.0320 | 29 | 0.0320 | 29 | 31.2416 | 31262 | 0.9995 | 1 | 10 | |
| 2 | 0 | 0.0349 | 29 | 0.0349 | 29 | 28.6363 | 26053 | 0.9994 | 1 | 0 | 88 |
| | 10 | 0.0378 | 29 | 0.0378 | 29 | 26.4316 | 22047 | 0.9993 | 1 | 50 | |
| | 20 | 0.0407 | 29 | 0.0407 | 29 | 24.5418 | 18898 | 0.9992 | 1 | 40 | |
| | 30 | 0.0436 | 29 | 0.0437 | 30 | 22.9038 | 16380 | 0.9990 | 2 | 30 | |
| | 40 | 0.0465 | 29 | 0.0466 | 29 | 21.4704 | 14334 | 0.9989 | 1 | 20 | |
| | 50 | 0.0494 | 29 | 0.0495 | 29 | 20.2056 | 12648 | 0.9988 | 1 | 10 | |
| 3 | 0 | 0.0523 | 29 | 0.0524 | 29 | 19.0811 | 11245 | 0.9986 | 2 | 0 | 87 |
| | 10 | 0.0552 | 29 | 0.0553 | 29 | 18.0750 | 10061 | 0.9985 | 1 | 50 | |
| | 20 | 0.0581 | 29 | 0.0582 | 29 | 17.1693 | 9057 | 0.9983 | 2 | 40 | |
| | 30 | 0.0610 | 29 | 0.0612 | 30 | 16.3499 | 8194 | 0.9981 | 2 | 30 | |
| | 40 | 0.0640 | 30 | 0.0641 | 29 | 15.6048 | 7451 | 0.9980 | 1 | 20 | |
| | 50 | 0.0669 | 29 | 0.0670 | 29 | 14.9244 | 6804 | 0.9978 | 2 | 10 | |
| 4 | 0 | 0.0698 | 29 | 0.0699 | 29 | 14.3007 | 6237 | 0.9976 | 2 | 0 | 86 |
| | 10 | 0.0727 | 29 | 0.0729 | 30 | 13.7267 | 5740 | 0.9974 | 2 | 50 | |
| | 20 | 0.0756 | 29 | 0.0758 | 29 | 13.1969 | 5298 | 0.9971 | 3 | 40 | |
| | 30 | 0.0785 | 29 | 0.0787 | 29 | 12.7062 | 4907 | 0.9969 | 2 | 30 | |
| | 40 | 0.0814 | 29 | 0.0816 | 29 | 12.2505 | 4557 | 0.9967 | 2 | 20 | |
| | 50 | 0.0843 | 29 | 0.0846 | 30 | 11.8262 | 4243 | 0.9964 | 3 | 10 | |
| 5 | 0 | 0.0872 | 29 | 0.0875 | 29 | 11.4301 | 3961 | 0.9962 | 2 | 0 | 85 |
| | 10 | 0.0901 | 29 | 0.0904 | 29 | 11.0594 | 3707 | 0.9959 | 3 | 50 | |
| | 20 | 0.0929 | 28 | 0.0934 | 30 | 10.7119 | 3475 | 0.9957 | 2 | 40 | |
| | 30 | 0.0958 | 29 | 0.0963 | 29 | 10.3854 | 3265 | 0.9954 | 3 | 30 | |
| | 40 | 0.0987 | 29 | 0.0992 | 29 | 10.0780 | 3074 | 0.9951 | 3 | 20 | |
| | 50 | 0.1016 | 29 | 0.1022 | 30 | 9.7882 | 2898 | 0.9948 | 3 | 10 | |
| 6 | 0 | 0.1045 | 29 | 0.1051 | 29 | 9.5144 | 2738 | 0.9945 | 3 | 0 | 84 |
| | 10 | 0.1074 | 29 | 0.1080 | 29 | 9.2553 | 2591 | 0.9942 | 3 | 50 | |
| | 20 | 0.1103 | 29 | 0.1110 | 30 | 9.0098 | 2455 | 0.9939 | 3 | 40 | |
| | 30 | 0.1132 | 29 | 0.1139 | 29 | 8.7769 | 2329 | 0.9936 | 3 | 30 | |
| | 40 | 0.1161 | 29 | 0.1169 | 30 | 8.5555 | 2214 | 0.9932 | 4 | 20 | |
| | 50 | 0.1190 | 29 | 0.1198 | 29 | 8.3450 | 2105 | 0.9929 | 3 | 10 | |
| 7 | 0 | 0.1219 | 29 | 0.1228 | 30 | 8.1443 | 2007 | 0.9925 | 4 | 0 | 83 |
| | 10 | 0.1248 | 29 | 0.1257 | 29 | 7.9530 | 1913 | 0.9922 | 3 | 50 | |
| | 20 | 0.1276 | 28 | 0.1287 | 30 | 7.7704 | 1826 | 0.9918 | 4 | 40 | |
| | 30 | 0.1305 | 29 | 0.1317 | 30 | 7.5958 | 1746 | 0.9914 | 4 | 30 | |
| | 40 | 0.1334 | 29 | 0.1346 | 29 | 7.4287 | 1671 | 0.9911 | 3 | 20 | |
| | 50 | 0.1363 | 29 | 0.1376 | 30 | 7.2687 | 1600 | 0.9907 | 4 | 10 | |
| 8 | 0 | 0.1392 | 29 | 0.1405 | 29 | 7.1154 | 1533 | 0.9903 | 4 | 0 | 82 |
| | | Cos. | d. | Cot. | d. | Tan. | d. | Sin. | d. | ′ | ° |

P. P.

| | 30 |
|---|---|
| 1 | 3.0 |
| 2 | 6.0 |
| 3 | 9.0 |
| 4 | 12.0 |
| 5 | 15.0 |
| 6 | 18.0 |
| 7 | 21.0 |
| 8 | 24.0 |
| 9 | 27.0 |

| | 29 |
|---|---|
| 1 | 2.9 |
| 2 | 5.8 |
| 3 | 8.7 |
| 4 | 11.6 |
| 5 | 14.5 |
| 6 | 17.4 |
| 7 | 20.3 |
| 8 | 23.2 |
| 9 | 26.1 |

| | 28 |
|---|---|
| 1 | 2.8 |
| 2 | 5.6 |
| 3 | 8.4 |
| 4 | 11.2 |
| 5 | 14.0 |
| 6 | 16.8 |
| 7 | 19.6 |
| 8 | 22.4 |
| 9 | 25.2 |

| | 5 |
|---|---|
| 1 | 0.5 |
| 2 | 1.0 |
| 3 | 1.5 |
| 4 | 2.0 |
| 5 | 2.5 |
| 6 | 3.0 |
| 7 | 3 5 |
| 8 | 4.0 |
| 9 | 4.5 |

| | 4 |
|---|---|
| 1 | 0.4 |
| 2 | 0.8 |
| 3 | 1.2 |
| 4 | 1.6 |
| 5 | 2.0 |
| 6 | 2.4 |
| 7 | 2.8 |
| 8 | 3.2 |
| 9 | 3.6 |

P. P.

## TRIGONOMETRIC FUNCTIONS. — *Continued.*

| ° | ′ | Sin. | d. | Tan. | d. | Cot. | d. | Cos. | d. | ′ | ° |
|---|---|---|---|---|---|---|---|---|---|---|---|
| **8** | 0 | 0.1392 | | 0.1405 | | 7.1154 | | 0.9903 | | 0 | **82** |
| | 10 | 0.1421 | 29 | 0.1435 | 30 | 6.9682 | 1472 | 0.9899 | 4 | 50 | |
| | 20 | 0.1449 | 28 | 0.1465 | 30 | 6.8269 | 1413 | 0.9894 | 5 | 40 | |
| | 30 | 0.1478 | 29 | 0.1495 | 30 | 6.6912 | 1357 | 0.9890 | 4 | 30 | |
| | 40 | 0.1507 | 29 | 0.1524 | 29 | 6.5606 | 1306 | 0.9886 | 4 | 20 | |
| | 50 | 0.1536 | 29 | 0.1554 | 30 | 6.4348 | 1258 | 0.9881 | 5 | 10 | |
| **9** | 0 | 0.1564 | 28 | 0.1584 | 30 | 6.3138 | 1210 | 0.9877 | 4 | 0 | **81** |
| | 10 | 0.1593 | 29 | 0.1614 | 30 | 6.1970 | 1168 | 0.9872 | 5 | 50 | |
| | 20 | 0.1622 | 29 | 0.1644 | 30 | 6.0844 | 1126 | 0.9868 | 4 | 40 | |
| | 30 | 0.1650 | 28 | 0.1673 | 29 | 5.9758 | 1086 | 0.9863 | 5 | 30 | |
| | 40 | 0.1679 | 29 | 0.1703 | 30 | 5.8708 | 1050 | 0.9858 | 5 | 20 | |
| | 50 | 0.1708 | 29 | 0.1733 | 30 | 5.7694 | 1014 | 0.9853 | 5 | 10 | |
| **10** | 0 | 0.1736 | 28 | 0.1763 | 30 | 5.6713 | 981 | 0.9848 | 5 | 0 | **80** |
| | 10 | 0.1765 | 29 | 0.1793 | 30 | 5.5764 | 949 | 0.9843 | 5 | 50 | |
| | 20 | 0.1794 | 29 | 0.1823 | 30 | 5.4845 | 919 | 0.9838 | 5 | 40 | |
| | 30 | 0.1822 | 28 | 0.1853 | 30 | 5.3955 | 890 | 0.9833 | 5 | 30 | |
| | 40 | 0.1851 | 29 | 0.1883 | 30 | 5.3093 | 862 | 0.9827 | 6 | 20 | |
| | 50 | 0.1880 | 29 | 0.1914 | 31 | 5.2257 | 836 | 0.9822 | 5 | 10 | |
| **11** | 0 | 0.1908 | 28 | 0.1944 | 30 | 5.1446 | 811 | 0.9816 | 6 | 0 | **79** |
| | 10 | 0.1937 | 29 | 0.1974 | 30 | 5.0658 | 788 | 0.9811 | 5 | 50 | |
| | 20 | 0.1965 | 28 | 0.2004 | 30 | 4.9894 | 764 | 0.9805 | 6 | 40 | |
| | 30 | 0.1994 | 29 | 0.2035 | 31 | 4.9152 | 742 | 0.9799 | 6 | 30 | |
| | 40 | 0.2022 | 28 | 0.2065 | 30 | 4.8430 | 722 | 0.9793 | 6 | 20 | |
| | 50 | 0.2051 | 29 | 0.2095 | 30 | 4.7729 | 701 | 0.9787 | 6 | 10 | |
| **12** | 0 | 0.2079 | 28 | 0.2126 | 31 | 4.7046 | 683 | 0.9781 | 6 | 0 | **78** |
| | 10 | 0.2108 | 29 | 0.2156 | 30 | 4.6382 | 664 | 0.9775 | 6 | 50 | |
| | 20 | 0.2136 | 28 | 0.2186 | 30 | 4.5736 | 646 | 0.9769 | 6 | 40 | |
| | 30 | 0.2164 | 28 | 0.2217 | 31 | 4.5107 | 629 | 0.9763 | 6 | 30 | |
| | 40 | 0.2193 | 29 | 0.2247 | 30 | 4.4494 | 613 | 0.9757 | 6 | 20 | |
| | 50 | 0.2221 | 28 | 0.2278 | 31 | 4.3897 | 597 | 0.9750 | 7 | 10 | |
| **13** | 0 | 0.2250 | 29 | 0.2309 | 31 | 4.3315 | 582 | 0.9744 | 6 | 0 | **77** |
| | 10 | 0.2278 | 28 | 0.2339 | 30 | 4.2747 | 568 | 0.9737 | 7 | 50 | |
| | 20 | 0.2306 | 28 | 0.2370 | 31 | 4.2193 | 554 | 0.9730 | 7 | 40 | |
| | 30 | 0.2334 | 28 | 0.2401 | 31 | 4.1653 | 540 | 0.9724 | 6 | 30 | |
| | 40 | 0.2363 | 29 | 0.2432 | 31 | 4.1126 | 527 | 0.9717 | 7 | 20 | |
| | 50 | 0.2391 | 28 | 0.2462 | 30 | 4.0611 | 515 | 0.9710 | 7 | 10 | |
| **14** | 0 | 0.2419 | 28 | 0.2493 | 31 | 4.0108 | 503 | 0.9703 | 7 | 0 | **76** |
| | 10 | 0.2447 | 28 | 0.2524 | 31 | 3.9617 | 491 | 0.9696 | 7 | 50 | |
| | 20 | 0.2476 | 29 | 0.2555 | 31 | 3.9136 | 481 | 0.9689 | 7 | 40 | |
| | 30 | 0.2504 | 28 | 0.2586 | 31 | 3.8667 | 469 | 0.9681 | 8 | 30 | |
| | 40 | 0.2532 | 28 | 0.2617 | 31 | 3.8208 | 459 | 0.9674 | 7 | 20 | |
| | 50 | 0.2560 | 28 | 0.2648 | 31 | 3.7760 | 448 | 0.9667 | 7 | 10 | |
| **15** | 0 | 0.2588 | 28 | 0.2679 | 31 | 3.7321 | 439 | 0.9659 | 8 | 0 | **75** |
| | 10 | 0.2616 | 28 | 0.2711 | 32 | 3.6891 | 430 | 0.9652 | 7 | 50 | |
| | 20 | 0.2644 | 28 | 0.2742 | 31 | 3.6470 | 421 | 0.9644 | 8 | 40 | |
| | 30 | 0.2672 | 28 | 0.2773 | 31 | 3.6059 | 411 | 0.9636 | 8 | 30 | |
| | 40 | 0.2700 | 28 | 0 2805 | 32 | 3.5656 | 403 | 0.9628 | 8 | 20 | |
| | 50 | 0.2728 | 28 | 0.2836 | 31 | 3.5261 | 395 | 0.9621 | 7 | 10 | |
| **16** | 0 | 0.2756 | 28 | 0.2867 | 31 | 3.4874 | 387 | 0.9613 | 8 | 0 | **74** |
| | | Cos. | d. | Cot. | d. | Tan. | d. | Sin. | d. | ′ | ° |

P. P.

| | 32 | 31 | 30 |
|---|---|---|---|
| **1** | 3.2 | 3.1 | 3.0 |
| **2** | 6.4 | 6.2 | 6.0 |
| **3** | 9.6 | 9.3 | 9.0 |
| **4** | 12.8 | 12.4 | 12.0 |
| **5** | 16.0 | 15.5 | 15.0 |
| **6** | 19.2 | 18.6 | 18.0 |
| **7** | 22.4 | 21.7 | 21.0 |
| **8** | 25.6 | 24.8 | 24.0 |
| **9** | 28.8 | 27.9 | 27.0 |

| | 29 | 28 | 27 |
|---|---|---|---|
| **1** | 2.9 | 2.8 | 2.7 |
| **2** | 5.8 | 5.6 | 5.4 |
| **3** | 8.7 | 8.4 | 8.1 |
| **4** | 11.6 | 11.2 | 10.8 |
| **5** | 14.5 | 14.0 | 13.5 |
| **6** | 17.4 | 16.8 | 16.2 |
| **7** | 20.3 | 19.6 | 18.9 |
| **8** | 23.2 | 22.4 | 21.6 |
| **9** | 26.1 | 25.2 | 24.3 |

| | 9 | 8 |
|---|---|---|
| **1** | 0.9 | 0.8 |
| **2** | 1.8 | 1.6 |
| **3** | 2.7 | 2.4 |
| **4** | 3.6 | 3.2 |
| **5** | 4.5 | 4.0 |
| **6** | 5.4 | 4.8 |
| **7** | 6.3 | 5.6 |
| **8** | 7.2 | 6.4 |
| **9** | 8.1 | 7.2 |

| | 7 | 6 |
|---|---|---|
| **1** | 0.7 | 0.6 |
| **2** | 1.4 | 1.2 |
| **3** | 2.1 | 1.8 |
| **4** | 2.8 | 2.4 |
| **5** | 3.5 | 3.0 |
| **6** | 4.2 | 3.6 |
| **7** | 4.9 | 4.2 |
| **8** | 5.6 | 4.8 |
| **9** | 6.3 | 5.4 |

P. P.

TRIGONOMETRIC FUNCTIONS. — *Continued.*

| ° | ′ | Sin. | d. | Tan. | d. | Cot. | d. | Cos. | d. | ′ | ° |
|---|---|---|---|---|---|---|---|---|---|---|---|
| 16 | 0 | 0.2756 | 28 | 0.2867 | 32 | 3.4874 | 379 | 0.9613 | 8 | 0 | 74 |
| | 10 | 0.2784 | 28 | 0.2899 | 32 | 3.4495 | 371 | 0.9605 | 9 | 50 | |
| | 20 | 0.2812 | 28 | 0.2931 | 31 | 3.4124 | 365 | 0.9596 | 8 | 40 | |
| | 30 | 0.2840 | 28 | 0.2962 | 32 | 3.3759 | 357 | 0.9588 | 8 | 30 | |
| | 40 | 0.2868 | 28 | 0.2994 | 32 | 3.3402 | 350 | 0.9580 | 8 | 20 | |
| | 50 | 0 2896 | 28 | 0.3026 | 31 | 3.3052 | 343 | 0.9572 | 9 | 10 | |
| 17 | 0 | 0.2924 | 28 | 0.3057 | 32 | 3.2709 | 338 | 0.9563 | 8 | 0 | 73 |
| | 10 | 0.2952 | 27 | 0.3089 | 32 | 3 2371 | 330 | 0.9555 | 9 | 50 | |
| | 20 | 0.2979 | 28 | 0.3121 | 32 | 3.2041 | 325 | 0.9546 | 9 | 40 | |
| | 30 | 0.3007 | 28 | 0.3153 | 32 | 3.1716 | 319 | 0.9537 | 9 | 30 | |
| | 40 | 0.3035 | 27 | 0.3185 | 32 | 3.1397 | 313 | 0.9528 | 8 | 20 | |
| | 50 | 0.3062 | 28 | 0.3217 | 32 | 3.1084 | 307 | 0.9520 | 9 | 10 | |
| 18 | 0 | 0.3090 | 28 | 0.3249 | 32 | 3.0777 | 302 | 0.9511 | 9 | 0 | 72 |
| | 10 | 0.3118 | 27 | 0.3281 | 33 | 3.0475 | 297 | 0.9502 | 10 | 50 | |
| | 20 | 0.3145 | 28 | 0.3314 | 32 | 3.0178 | 291 | 0.9492 | 9 | 40 | |
| | 30 | 0.3173 | 28 | 0.3346 | 32 | 2.9887 | 287 | 0.9483 | 9 | 30 | |
| | 40 | 0.3201 | 27 | 0.3378 | 33 | 2.9600 | 281 | 0.9474 | 9 | 20 | |
| | 50 | 0.3228 | 28 | 0.3411 | 32 | 2.9319 | 277 | 0.9465 | 10 | 10 | |
| 19 | 0 | 0.3256 | 27 | 0.3443 | 33 | 2.9042 | 272 | 0.9455 | 9 | 0 | 71 |
| | 10 | 0.3283 | 28 | 0.3476 | 32 | 2.8770 | 268 | 0.9446 | 10 | 50 | |
| | 20 | 0.3311 | 27 | 0.3508 | 33 | 2.8502 | 263 | 0.9436 | 10 | 40 | |
| | 30 | 0.3338 | 27 | 0.3541 | 33 | 2.8239 | 259 | 0.9426 | 9 | 30 | |
| | 40 | 0.3365 | 28 | 0.3574 | 33 | 2 7980 | 255 | 0.9417 | 10 | 20 | |
| | 50 | 0.3393 | 27 | 0.3607 | 33 | 2 7725 | 250 | 0.9407 | 10 | 10 | |
| 20 | 0 | 0.3420 | 28 | 0.3640 | 33 | 2.7475 | 247 | 0.9397 | 10 | 0 | 70 |
| | 10 | 0.3448 | 27 | 0.3673 | 33 | 2.7228 | 243 | 0.9387 | 10 | 50 | |
| | 20 | 0.3475 | 27 | 0.3706 | 33 | 2.6985 | 239 | 0.9377 | 10 | 40 | |
| | 30 | 0.3502 | 27 | 0.3739 | 33 | 2 6746 | 235 | 0.9367 | 11 | 30 | |
| | 40 | 0.3529 | 28 | 0.3772 | 33 | 2.6511 | 232 | 0.9356 | 10 | 20 | |
| | 50 | 0.3557 | 27 | 0.3805 | 34 | 2.6279 | 228 | 0.9346 | 10 | 10 | |
| 21 | 0 | 0.3584 | 27 | 0.3839 | 33 | 2.6051 | 225 | 0.9336 | 11 | 0 | 69 |
| | 10 | 0.3611 | 27 | 0.3872 | 34 | 2.5826 | 221 | 0.9325 | 10 | 50 | |
| | 20 | 0.3638 | 27 | 0.3906 | 33 | 2.5605 | 219 | 0.9315 | 11 | 40 | |
| | 30 | 0.3665 | 27 | 0.3939 | 34 | 2.5386 | 214 | 0.9304 | 11 | 30 | |
| | 40 | 0.3692 | 27 | 0.3973 | 33 | 2.5172 | 212 | 0.9293 | 10 | 20 | |
| | 50 | 0.3719 | 27 | 0.4006 | 34 | 2.4960 | 209 | 0.9283 | 11 | 10 | |
| 22 | 0 | 0.3746 | 27 | 0.4040 | 34 | 2.4751 | 206 | 0.9272 | 11 | 0 | 68 |
| | 10 | 0.3773 | 27 | 0.4074 | 34 | 2.4545 | 203 | 0.9261 | 11 | 50 | |
| | 20 | 0.3800 | 27 | 0.4108 | 34 | 2.4342 | 200 | 0.9250 | 11 | 40 | |
| | 30 | 0.3827 | 27 | 0.4142 | 34 | 2.4142 | 197 | 0.9239 | 11 | 30 | |
| | 40 | 0.3854 | 27 | 0.4176 | 34 | 2.3945 | 195 | 0.9228 | 12 | 20 | |
| | 50 | 0.3881 | 26 | 0.4210 | 35 | 2.3750 | 191 | 0.9216 | 11 | 10 | |
| 23 | 0 | 0.3907 | 27 | 0.4245 | 34 | 2.3559 | 190 | 0.9205 | 11 | 0 | 67 |
| | 10 | 0.3934 | 27 | 0.4279 | 35 | 2.3369 | 186 | 0.9194 | 12 | 50 | |
| | 20 | 0.3961 | 26 | 0.4314 | 34 | 2.3183 | 185 | 0.9182 | 11 | 40 | |
| | 30 | 0.3987 | 27 | 0.4348 | 35 | 2.2998 | 181 | 0.9171 | 12 | 30 | |
| | 40 | 0.4014 | 27 | 0.4383 | 34 | 2.2817 | 180 | 0.9159 | 12 | 20 | |
| | 50 | 0.4041 | 26 | 0.4417 | 35 | 2.2637 | 177 | 0.9147 | 12 | 10 | |
| 24 | | 0.4067 | | 0.4452 | | 2.2460 | 174 | 0.9135 | | 0 | 66 |
| | | Cos. | d. | Cot. | d. | Tan. | d. | Sin. | d. | ′ | ° |

P. P.

| | 5 | 4 |
|---|---|---|
| 1 | 0.5 | 0.4 |
| 2 | 1.0 | 0.8 |
| 3 | 1.5 | 1.2 |
| 4 | 2.0 | 1.6 |
| 5 | 2.5 | 2.0 |
| 6 | 3.0 | 2.4 |
| 7 | 3.5 | 2.8 |
| 8 | 4.0 | 3.2 |
| 9 | 4.5 | 3.6 |

| | 37 | 36 | 35 |
|---|---|---|---|
| 1 | 3.7 | 3.6 | 3.5 |
| 2 | 7.4 | 7.2 | 7.0 |
| 3 | 11.1 | 10.8 | 10.5 |
| 4 | 14.8 | 14.4 | 14.0 |
| 5 | 18.5 | 18.0 | 17.5 |
| 6 | 22.2 | 21.6 | 21.0 |
| 7 | 25.9 | 25.2 | 24.5 |
| 8 | 29.6 | 28.8 | 28.0 |
| 9 | 33.3 | 32.4 | 31.5 |

| | 34 | 33 | 32 |
|---|---|---|---|
| 1 | 3.4 | 3.3 | 3.2 |
| 2 | 6.8 | 6.6 | 6.4 |
| 3 | 10.2 | 9.9 | 9.6 |
| 4 | 13.6 | 13 2 | 12.8 |
| 5 | 17.0 | 16.5 | 16.0 |
| 6 | 20.4 | 19.8 | 19.2 |
| 7 | 23.8 | 23.1 | 22.4 |
| 8 | 27.2 | 26.4 | 25.6 |
| 9 | 30.6 | 29.7 | 28.8 |

| | 28 | 27 | 26 |
|---|---|---|---|
| 1 | 2.8 | 2.7 | 2.6 |
| 2 | 5.6 | 5.4 | 5.2 |
| 3 | 8.4 | 8.1 | 7.8 |
| 4 | 11.2 | 10.8 | 10.4 |
| 5 | 14.0 | 13.5 | 13.0 |
| 6 | 16.8 | 16.2 | 15.6 |
| 7 | 19.6 | 18.9 | 18.2 |
| 8 | 22.4 | 21.6 | 20.8 |
| 9 | 25.2 | 24.3 | 23.4 |

| | 13 | 12 |
|---|---|---|
| 1 | 1.3 | 1.2 |
| 2 | 2.6 | 2.4 |
| 3 | 3.9 | 3.6 |
| 4 | 5.2 | 4.8 |
| 5 | 6.5 | 6.0 |
| 6 | 7.8 | 7.2 |
| 7 | 9.1 | 8.4 |
| 8 | 10.4 | 9.6 |
| 9 | 11.7 | 10.8 |

P. P.

## TRIGONOMETRIC FUNCTIONS. — *Continued.*

| ° | ′ | Sin. | d. | Tan. | d. | Cot. | d. | Cos. | d. | ′ | ° |
|---|---|---|---|---|---|---|---|---|---|---|---|
| **24** | 0 | 0.4067 | 27 | 0.4452 | 35 | 2.2460 | 174 | 0.9135 | 11 | 0 | **66** |
| | 10 | 0.4094 | 26 | 0.4487 | 35 | 2.2286 | 173 | 0.9124 | 12 | 50 | |
| | 20 | 0.4120 | 27 | 0.4522 | 35 | 2.2113 | 170 | 0.9112 | 12 | 40 | |
| | 30 | 0.4147 | 26 | 0.4557 | 35 | 2.1943 | 168 | 0.9100 | 12 | 30 | |
| | 40 | 0.4173 | 27 | 0.4592 | 36 | 2.1775 | 166 | 0.9088 | 13 | 20 | |
| | 50 | 0.4200 | 26 | 0.4628 | 35 | 2.1609 | 164 | 0.9075 | 12 | 10 | |
| **25** | 0 | 0.4226 | 27 | 0.4663 | 36 | 2.1445 | 162 | 0.9063 | 12 | 0 | **65** |
| | 10 | 0.4253 | 26 | 0.4699 | 35 | 2.1283 | 160 | 0.9051 | 13 | 50 | |
| | 20 | 0.4279 | 26 | 0.4734 | 36 | 2.1123 | 158 | 0.9038 | 12 | 40 | |
| | 30 | 0.4305 | 26 | 0.4770 | 36 | 2.0965 | 156 | 0.9026 | 13 | 30 | |
| | 40 | 0.4331 | 27 | 0.4806 | 35 | 2.0809 | 154 | 0.9013 | 12 | 20 | |
| | 50 | 0.4358 | 26 | 0.4841 | 36 | 2.0655 | 152 | 0.9001 | 13 | 10 | |
| **26** | 0 | 0.4384 | 26 | 0.4877 | 36 | 2.0503 | 150 | 0.8988 | 13 | 0 | **64** |
| | 10 | 0.4410 | 26 | 0.4913 | 37 | 2.0353 | 149 | 0.8975 | 13 | 50 | |
| | 20 | 0.4436 | 26 | 0.4950 | 36 | 2.0204 | 147 | 0.8962 | 13 | 40 | |
| | 30 | 0.4462 | 26 | 0.4986 | 36 | 2.0057 | 145 | 0.8949 | 13 | 30 | |
| | 40 | 0.4488 | 26 | 0.5022 | 37 | 1.9912 | 144 | 0.8936 | 13 | 20 | |
| | 50 | 0.4514 | 26 | 0.5059 | 36 | 1.9768 | 142 | 0.8923 | 13 | 10 | |
| **27** | 0 | 0.4540 | 26 | 0.5095 | 37 | 1.9626 | 140 | 0.8910 | 13 | 0 | **63** |
| | 10 | 0.4566 | 26 | 0.5132 | 37 | 1.9486 | 139 | 0.8897 | 13 | 50 | |
| | 20 | 0.4592 | 25 | 0.5169 | 37 | 1.9347 | 137 | 0.8884 | 14 | 40 | |
| | 30 | 0.4617 | 26 | 0.5206 | 37 | 1.9210 | 136 | 0.8870 | 13 | 30 | |
| | 40 | 0.4643 | 26 | 0.5243 | 37 | 1.9074 | 134 | 0.8857 | 14 | 20 | |
| | 50 | 0.4669 | 26 | 0.5280 | 37 | 1.8940 | 133 | 0.8843 | 14 | 10 | |
| **28** | 0 | 0.4695 | 25 | 0.5317 | 37 | 1.8807 | 131 | 0.8829 | 13 | 0 | **62** |
| | 10 | 0.4720 | 26 | 0.5354 | 38 | 1.8676 | 130 | 0.8816 | 14 | 50 | |
| | 20 | 0.4746 | 26 | 0.5392 | 38 | 1.8546 | 128 | 0.8802 | 14 | 40 | |
| | 30 | 0.4772 | 25 | 0.5430 | 37 | 1.8418 | 127 | 0.8788 | 14 | 30 | |
| | 40 | 0.4797 | 26 | 0.5467 | 38 | 1.8291 | 126 | 0.8774 | 14 | 20 | |
| | 50 | 0.4823 | 25 | 0.5505 | 38 | 1.8165 | 125 | 0.8760 | 14 | 10 | |
| **29** | 0 | 0.4848 | 26 | 0.5543 | 38 | 1.8040 | 123 | 0.8746 | 14 | 0 | **61** |
| | 10 | 0.4874 | 25 | 0.5581 | 38 | 1.7917 | 121 | 0.8732 | 14 | 50 | |
| | 20 | 0.4899 | 25 | 0.5619 | 39 | 1.7796 | 121 | 0.8718 | 14 | 40 | |
| | 30 | 0.4924 | 26 | 0.5658 | 38 | 1.7675 | 119 | 0.8704 | 15 | 30 | |
| | 40 | 0.4950 | 25 | 0.5696 | 39 | 1.7556 | 119 | 0.8689 | 14 | 20 | |
| | 50 | 0.4975 | 25 | 0.5735 | 39 | 1.7437 | 116 | 0.8675 | 15 | 10 | |
| **30** | 0 | 0.5000 | 25 | 0.5774 | 38 | 1.7321 | 116 | 0 8660 | 14 | 0 | **60** |
| | 10 | 0.5025 | 25 | 0.5812 | 39 | 1.7205 | 115 | 0.8646 | 15 | 50 | |
| | 20 | 0.5050 | 25 | 0.5851 | 39 | 1.7090 | 113 | 0.8631 | 15 | 40 | |
| | 30 | 0.5075 | 25 | 0.5890 | 40 | 1.6977 | 113 | 0.8616 | 15 | 30 | |
| | 40 | 0.5100 | 25 | 0.5930 | 39 | 1.6864 | 111 | 0.8601 | 14 | 20 | |
| | 50 | 0.5125 | 25 | 0.5969 | 40 | 1.6753 | 110 | 0.8587 | 15 | 10 | |
| **31** | 0 | 0.5150 | 25 | 0.6009 | 39 | 1.6643 | 109 | 0.8572 | 15 | 0 | **59** |
| | 10 | 0.5175 | 25 | 0.6048 | 40 | 1.6534 | 108 | 0.8557 | 15 | 50 | |
| | 20 | 0.5200 | 25 | 0.6088 | 40 | 1.6426 | 107 | 0.8542 | 16 | 40 | |
| | 30 | 0.5225 | 25 | 0.6128 | 40 | 1.6319 | 107 | 0.8526 | 15 | 30 | |
| | 40 | 0.5250 | 25 | 0.6168 | 40 | 1.6212 | 105 | 0.8511 | 15 | 20 | |
| | 50 | 0.5275 | 24 | 0.6208 | 41 | 1.6107 | 104 | 0.8496 | 16 | 10 | |
| **32** | 0 | 5299 | | 6249 | | 1.6003 | | 0.8480 | | 0 | **58** |
| | | Cos. | d | Cot. | d. | Tan. | d. | Sin. | d. | ′ | ° |

P. P.

| | 11 | 10 | 9 |
|---|---|---|---|
| 1 | 1.1 | 1.0 | 0.9 |
| 2 | 2.2 | 2.0 | 1.8 |
| 3 | 3.3 | 3.0 | 2.7 |
| 4 | 4.4 | 4.0 | 3.6 |
| 5 | 5.5 | 5.0 | 4.5 |
| 6 | 6.6 | 6.0 | 5.4 |
| 7 | 7.7 | 7.0 | 6.3 |
| 8 | 8.8 | 8.0 | 7.2 |
| 9 | 9.9 | 9.0 | 8.0 |

| | 44 | 43 | 42 |
|---|---|---|---|
| 1 | 4.4 | 4.3 | 4.2 |
| 2 | 8.8 | 8.6 | 8.4 |
| 3 | 13.2 | 12.9 | 12.6 |
| 4 | 17.6 | 17.2 | 16.8 |
| 5 | 22.0 | 21.5 | 21.0 |
| 6 | 26.4 | 25.8 | 25.2 |
| 7 | 30.8 | 30.1 | 29.4 |
| 8 | 35.2 | 34.4 | 33.6 |
| 9 | 39.6 | 38.7 | 37.8 |

| | 41 | 40 | 39 |
|---|---|---|---|
| 1 | 4.1 | 4.0 | 3.9 |
| 2 | 8.2 | 8.0 | 7.8 |
| 3 | 12.3 | 12.0 | 11.7 |
| 4 | 16.4 | 16.0 | 15.6 |
| 5 | 20.5 | 20.0 | 19.5 |
| 6 | 24.6 | 24.0 | 23.4 |
| 7 | 28.7 | 28.0 | 27.3 |
| 8 | 32.8 | 32.0 | 31.2 |
| 9 | 36.9 | 36.0 | 35.1 |

| | 38 | 37 |
|---|---|---|
| 1 | 3.8 | 3.7 |
| 2 | 7.6 | 7.4 |
| 3 | 11.4 | 11.1 |
| 4 | 15.2 | 14.8 |
| 5 | 19.0 | 18.5 |
| 6 | 22.8 | 22.2 |
| 7 | 26.6 | 25.9 |
| 8 | 30.4 | 29.6 |
| 9 | 34.2 | 33.3 |

| | 26 | 25 | 24 |
|---|---|---|---|
| 1 | 2.6 | 2.5 | 2.4 |
| 2 | 5.2 | 5.0 | 4.8 |
| 3 | 7.8 | 7.5 | 7.2 |
| 4 | 10.4 | 10.0 | 9.6 |
| 5 | 13.0 | 12.5 | 12.0 |
| 6 | 15.6 | 15.0 | 14.4 |
| 7 | 18.2 | 17.5 | 16.8 |
| 8 | 20.8 | 20.0 | 19.2 |
| 9 | 23.4 | 22.5 | 21.6 |

P. P.

TRIGONOMETRIC FUNCTIONS.—*Continued.*

| ° | ′ | Sin. | d. | Tan. | d. | Cot. | d. | Cos. | d. | ′ | ° |
|---|---|---|---|---|---|---|---|---|---|---|---|
| **32** | 0 | 0.5299 | 25 | 0.6249 | 40 | 1.6003 | 103 | 0.8480 | 15 | 0 | **58** |
| | 10 | 0.5324 | 24 | 0.6289 | 41 | 1.5900 | 102 | 0.8465 | 15 | 50 | |
| | 20 | 0.5348 | 25 | 0.6330 | 41 | 1.5798 | 101 | 0.8450 | 16 | 40 | |
| | 30 | 0.5373 | 25 | 0.6371 | 41 | 1.5697 | 100 | 0.8434 | 16 | 30 | |
| | 40 | 0.5398 | 24 | 0.6412 | 41 | 1.5597 | 100 | 0.8418 | 15 | 20 | |
| | 50 | 0.5422 | 24 | 0.6453 | 41 | 1.5497 | 98 | 0.8403 | 16 | 10 | |
| **33** | 0 | 0.5446 | 25 | 0.6494 | 42 | 1.5399 | 98 | 0.8387 | 16 | 0 | **57** |
| | 10 | 0.5471 | 24 | 0.6536 | 41 | 1.5301 | 97 | 0.8371 | 16 | 50 | |
| | 20 | 0.5495 | 24 | 0.6577 | 42 | 1.5204 | 96 | 0.8355 | 16 | 40 | |
| | 30 | 0.5519 | 25 | 0.6619 | 42 | 1.5108 | 95 | 0.8339 | 16 | 30 | |
| | 40 | 0.5544 | 24 | 0.6661 | 42 | 1.5013 | 94 | 0.8323 | 16 | 20 | |
| | 50 | 0.5568 | 24 | 0.6703 | 42 | 1.4919 | 93 | 0.8307 | 17 | 10 | |
| **34** | 0 | 0.5592 | 24 | 0.6745 | 42 | 1.4826 | 93 | 0.8290 | 16 | 0 | **56** |
| | 10 | 0.5616 | 24 | 0.6787 | 43 | 1.4733 | 92 | 0.8274 | 16 | 50 | |
| | 20 | 0.5640 | 24 | 0.6830 | 43 | 1.4641 | 91 | 0.8258 | 17 | 40 | |
| | 30 | 0.5664 | 24 | 0.6873 | 43 | 1.4550 | 90 | 0.8241 | 16 | 30 | |
| | 40 | 0.5688 | 24 | 0.6916 | 43 | 1.4460 | 90 | 0.8225 | 17 | 20 | |
| | 50 | 0.5712 | 24 | 0.6959 | 43 | 1.4370 | 89 | 0.8208 | 16 | 10 | |
| **35** | 0 | 0.5736 | 24 | 0.7002 | 44 | 1.4281 | 88 | 0.8192 | 17 | 0 | **55** |
| | 10 | 0.5760 | 23 | 0.7046 | 43 | 1.4193 | 87 | 0.8175 | 17 | 50 | |
| | 20 | 0.5783 | 24 | 0.7089 | 44 | 1.4106 | 87 | 0.8158 | 17 | 40 | |
| | 30 | 0.5807 | 24 | 0.7133 | 44 | 1.4019 | 85 | 0.8141 | 17 | 30 | |
| | 40 | 0.5831 | 23 | 0.7177 | 44 | 1 3934 | 86 | 0.8124 | 17 | 20 | |
| | 50 | 0.5854 | 24 | 0 7221 | 44 | 1.3848 | 84 | 0.8107 | 17 | 10 | |
| **36** | 0 | 0.5878 | 23 | 0.7265 | 45 | 1.3764 | 84 | 0.8090 | 17 | 0 | **54** |
| | 10 | 0.5901 | 24 | 0.7310 | 45 | 1.3680 | 83 | 0.8073 | 17 | 50 | |
| | 20 | 0.5925 | 23 | 0.7355 | 45 | 1.3597 | 83 | 0.8056 | 17 | 40 | |
| | 30 | 0.5948 | 24 | 0.7400 | 45 | 1.3514 | 82 | 0.8039 | 18 | 30 | |
| | 40 | 0.5972 | 23 | 0.7445 | 45 | 1.3432 | 81 | 0.8021 | 17 | 20 | |
| | 50 | 0.5995 | 23 | 0.7490 | 46 | 1.3351 | 81 | 0.8004 | 18 | 10 | |
| **37** | 0 | 0.6018 | 23 | 0.7536 | 45 | 1.3270 | 80 | 0.7986 | 17 | 0 | **53** |
| | 10 | 0.6041 | 24 | 0.7581 | 46 | 1.3190 | 79 | 0.7969 | 18 | 50 | |
| | 20 | 0.6065 | 23 | 0.7627 | 46 | 1.3111 | 79 | 0.7951 | 17 | 40 | |
| | 30 | 0.6088 | 23 | 0.7673 | 47 | 1.3032 | 78 | 0.7934 | 18 | 30 | |
| | 40 | 0.6111 | 23 | 0.7720 | 46 | 1.2954 | 78 | 0.7916 | 18 | 20 | |
| | 50 | 0.6134 | 23 | 0.7766 | 47 | 1.2876 | 77 | 0.7898 | 18 | 10 | |
| **38** | 0 | 0.6157 | 23 | 0.7813 | 47 | 1.2799 | 76 | 0.7880 | 18 | 0 | **52** |
| | 10 | 0.6180 | 22 | 0.7860 | 47 | 1.2723 | 76 | 0.7862 | 18 | 50 | |
| | 20 | 0.6202 | 23 | 0.7907 | 47 | 1.2647 | 75 | 0.7844 | 18 | 40 | |
| | 30 | 0.6225 | 23 | 0.7954 | 48 | 1.2572 | 75 | 0.7826 | 18 | 30 | |
| | 40 | 0.6248 | 23 | 0.8002 | 48 | 1.2497 | 74 | 0.7808 | 18 | 20 | |
| | 50 | 0.6271 | 22 | 0.8050 | 48 | 1.2423 | 74 | 0.7790 | 19 | 10 | |
| **39** | 0 | 0.6293 | 23 | 0.8098 | 48 | 1.2349 | 73 | 0.7771 | 18 | 0 | **51** |
| | 10 | 0.6316 | 22 | 0.8146 | 49 | 1.2276 | 73 | 0.7753 | 18 | 50 | |
| | 20 | 0.6338 | 23 | 0.8195 | 48 | 1.2203 | 72 | 0.7735 | 19 | 40 | |
| | 30 | 0.6361 | 22 | 0.8243 | 49 | 1.2131 | 72 | 0.7716 | 18 | 30 | |
| | 40 | 0.6383 | 23 | 0.8292 | 50 | 1.2059 | 71 | 0.7698 | 19 | 20 | |
| | 50 | 0.6406 | 22 | 0 8342 | 49 | 1.1988 | 70 | 0.7679 | 19 | 10 | |
| **40** | 0 | 0.6428 | | 0.8391 | | 1.1918 | | 0.7660 | | 0 | **50** |
| | | Cos. | d. | Cot. | d. | Tan. | d. | Sin. | d. | ′ | ° |

P. P.

| | **23** | **17** | **16** |
|---|---|---|---|
| **1** | 2.3 | 1.7 | 1.6 |
| **2** | 4.6 | 3.4 | 3.2 |
| **3** | 6.9 | 5.1 | 4.8 |
| **4** | 9.2 | 6.8 | 6.4 |
| **5** | 11.5 | 8.5 | 8.0 |
| **6** | 13.8 | 10.2 | 9.6 |
| **7** | 16.1 | 11.9 | 11.2 |
| **8** | 18.4 | 13.6 | 12.8 |
| **9** | 20.7 | 15.3 | 14.4 |

| | **15** | **14** | **13** |
|---|---|---|---|
| **1** | 1.5 | 1.4 | 1.3 |
| **2** | 3.0 | 2.8 | 2.6 |
| **3** | 4.5 | 4.2 | 3.9 |
| **4** | 6.0 | 5.6 | 5.2 |
| **5** | 7.5 | 7.0 | 6.5 |
| **6** | 9.0 | 8.4 | 7.8 |
| **7** | 10.5 | 9.8 | 9.1 |
| **8** | 12.0 | 11.2 | 10.4 |
| **9** | 13.5 | 12.6 | 11.7 |

| | **58** | **57** | **56** | **55** |
|---|---|---|---|---|
| **1** | 5.8 | 5 7 | 5.6 | 5 5 |
| **2** | 11.6 | 11.4 | 11.2 | 11.0 |
| **3** | 17.4 | 17.1 | 16.8 | 16.5 |
| **4** | 23.2 | 22.8 | 22.4 | 22.0 |
| **5** | 29.0 | 28.5 | 28.0 | 27.5 |
| **6** | 34.8 | 34.2 | 33.6 | 33.0 |
| **7** | 40.6 | 39.9 | 39.2 | 38.5 |
| **8** | 46.4 | 45.6 | 44.8 | 44.0 |
| **9** | 52.2 | 51.3 | 50.4 | 49.5 |

| | **54** | **53** | **52** | **51** |
|---|---|---|---|---|
| **1** | 5.4 | 5.3 | 5.2 | 5.1 |
| **2** | 10.8 | 10.6 | 10.4 | 10.2 |
| **3** | 16.2 | 15.9 | 15.6 | 15.3 |
| **4** | 21.6 | 21.2 | 20.8 | 20.4 |
| **5** | 27.0 | 26.5 | 26.0 | 25.5 |
| **6** | 32.4 | 31.8 | 31.2 | 30.6 |
| **7** | 37.8 | 37.1 | 36.4 | 35.7 |
| **8** | 43.2 | 42.4 | 41.6 | 40.8 |
| **9** | 48.6 | 47.7 | 46.8 | 45.9 |

| | **50** | **49** | **48** |
|---|---|---|---|
| **1** | 5.0 | 4.9 | 4.8 |
| **2** | 10.0 | 9.8 | 9.6 |
| **3** | 15.0 | 14.7 | 14.4 |
| **4** | 20.0 | 19.6 | 19.2 |
| **5** | 25.0 | 24.5 | 24.0 |
| **6** | 30.0 | 29.4 | 28.8 |
| **7** | 35.0 | 34.3 | 33.6 |
| **8** | 40.0 | 39.2 | 38.4 |
| **9** | 45.0 | 44.1 | 43.2 |

P. P.

## TRIGONOMETRIC FUNCTIONS.— *Continued.*

| ° | ′ | Sin. | d. | Tan. | d. | Cot. | d. | Cos. | d. | ′ ° |
|---|---|---|---|---|---|---|---|---|---|---|
| **40** | 0 | 0.6428 | | 0.8391 | | 1.1918 | | 0.7660 | | 0 **50** |
| | 10 | 0.6450 | 22 | 0.8441 | 50 | 1.1847 | 71 | 0.7642 | 18 | 50 |
| | 20 | 0.6472 | 22 | 0.8491 | 50 | 1.1778 | 69 | 0.7623 | 19 | 40 |
| | 30 | 0.6494 | 22 | 0.8541 | 50 | 1.1708 | 70 | 0.7604 | 19 | 30 |
| | 40 | 0.6517 | 23 | 0.8591 | 50 | 1.1640 | 68 | 0.7585 | 19 | 20 |
| | 50 | 0.6539 | 22 | 0.8642 | 51 | 1.1571 | 69 | 0.7566 | 19 | 10 |
| **41** | 0 | 0.6561 | 22 | 0.8693 | 51 | 1.1504 | 67 | 0.7547 | 19 | 0 **49** |
| | 10 | 0.6583 | 22 | 0.8744 | 51 | 1.1436 | 68 | 0.7528 | 19 | 50 |
| | 20 | 0.6604 | 21 | 0.8796 | 52 | 1.1369 | 67 | 0.7509 | 19 | 40 |
| | 30 | 0.6626 | 22 | 0.8847 | 51 | 1.1303 | 66 | 0.7490 | 19 | 30 |
| | 40 | 0.6648 | 22 | 0.8899 | 52 | 1.1237 | 66 | 0.7470 | 20 | 20 |
| | 50 | 0.6670 | 22 | 0.8952 | 53 | 1.1171 | 66 | 0.7451 | 19 | 10 |
| **42** | 0 | 0.6691 | 21 | 0.9004 | 52 | 1.1106 | 65 | 0.7431 | 20 | 0 **48** |
| | 10 | 0.6713 | 22 | 0.9057 | 53 | 1.1041 | 65 | 0.7412 | 19 | 50 |
| | 20 | 0.6734 | 21 | 0.9110 | 53 | 1.0977 | 64 | 0.7392 | 20 | 40 |
| | 30 | 0.6756 | 22 | 0.9163 | 53 | 1.0913 | 64 | 0.7373 | 19 | 30 |
| | 40 | 0.6777 | 21 | 0.9217 | 54 | 1.0850 | 63 | 0.7353 | 20 | 20 |
| | 50 | 0.6799 | 22 | 0.9271 | 54 | 1.0786 | 64 | 0.7333 | 20 | 10 |
| **43** | 0 | 0.6820 | 21 | 0.9325 | 54 | 1.0724 | 62 | 0.7314 | 19 | 0 **47** |
| | 10 | 0.6841 | 21 | 0.9380 | 55 | 1.0661 | 63 | 0.7294 | 20 | 50 |
| | 20 | 0.6862 | 21 | 0.9435 | 55 | 1.0599 | 62 | 0.7274 | 20 | 40 |
| | 30 | 0.6884 | 22 | 0.9490 | 55 | 1.0538 | 61 | 0.7254 | 20 | 30 |
| | 40 | 0.6905 | 21 | 0.9545 | 55 | 1.0477 | 61 | 0.7234 | 20 | 20 |
| | 50 | 0.6926 | 21 | 0.9601 | 56 | 1.0416 | 61 | 0.7214 | 20 | 10 |
| **44** | 0 | 0.6947 | 21 | 0.9657 | 56 | 1.0355 | 61 | 0.7193 | 21 | 0 **46** |
| | 10 | 0.6967 | 20 | 0.9713 | 56 | 1.0295 | 60 | 0.7173 | 20 | 50 |
| | 20 | 0.6988 | 21 | 0.9770 | 57 | 1.0235 | 60 | 0.7153 | 20 | 40 |
| | 30 | 0.7009 | 21 | 0.9827 | 57 | 1.0176 | 59 | 0.7133 | 20 | 30 |
| | 40 | 0.7030 | 21 | 0.9884 | 57 | 1.0117 | 59 | 0.7112 | 21 | 20 |
| | 50 | 0.7050 | 20 | 0.9942 | 58 | 1.0058 | 59 | 0.7092 | 20 | 10 |
| **45** | 0 | 0.7071 | 21 | 1.0000 | 58 | 1.0000 | 58 | 0.7071 | 21 | 0 **45** |
| | | Cos. | d. | Cot. | d. | Tan. | d. | Sin. | d. | ′ ° |

**P. P.**

| | 47 | 46 | 45 |
|---|---|---|---|
| 1 | 4.7 | 4.6 | 4.5 |
| 2 | 9.4 | 9.2 | 9.0 |
| 3 | 14.1 | 13.8 | 13.5 |
| 4 | 18.8 | 18.4 | 18 0 |
| 5 | 23.5 | 23.0 | 22.5 |
| 6 | 28.2 | 27.6 | 27.0 |
| 7 | 32.9 | 32.2 | 31.5 |
| 8 | 37.6 | 36.8 | 36.0 |
| 9 | 42.3 | 41.4 | 40.5 |

| | 24 | 23 | 22 | 21 |
|---|---|---|---|---|
| 1 | 2.4 | 2.3 | 2.2 | 2.1 |
| 2 | 4.8 | 4.6 | 4.4 | 4.2 |
| 3 | 7.2 | 6.9 | 6.6 | 6.3 |
| 4 | 9.6 | 9.2 | 8.8 | 8.4 |
| 5 | 12.0 | 11.5 | 11.0 | 10.5 |
| 6 | 14.4 | 13.8 | 13.2 | 12.6 |
| 7 | 16.8 | 16.1 | 15.4 | 14.7 |
| 8 | 19.2 | 18.4 | 17.6 | 16.8 |
| 9 | 21.6 | 20.7 | 19.8 | 18.9 |

| | 20 | 19 | 18 | 17 |
|---|---|---|---|---|
| 1 | 2.0 | 1.9 | 1.8 | 1.7 |
| 2 | 4.0 | 3.8 | 3.6 | 3.4 |
| 3 | 6.0 | 5.7 | 5.4 | 5.1 |
| 4 | 8.0 | 7.6 | 7.2 | 6.8 |
| 5 | 10.0 | 9.5 | 9.0 | 8.5 |
| 6 | 12.0 | 11.4 | 10.8 | 10.2 |
| 7 | 14.0 | 13.3 | 12.6 | 11.9 |
| 8 | 16.0 | 15.2 | 14.4 | 13.6 |
| 9 | 18.0 | 17.1 | 16.2 | 15.3 |

P. P.

# TABLE XXI.

# TABLE

OF

# COMMON LOGARITHMS

OF

# NUMBERS

FROM 1000 TO 10000

A *minus* sign (−) before or after any tabular log. indicates that its true value is less than the tabulated value by less than half of the unit in the last place.

## LOGARITHMS OF NUMBERS.

| No. | Log. | Dif. | No. | Log. | Dif. | No. | Log. | Dif. | No. | Log. | Dif. |
|---|---|---|---|---|---|---|---|---|---|---|---|
| **1000** | 00000 | 43 | **1050** | 02119 | 41 | **1100** | 04139 | 40 | **1150** | 06070 | 38 |
| **01** | 043 | 44 | **51** | 160 | 42 | **01** | −179 | 39 | **51** | −108 | 37 |
| **02** | −087 | 43 | **52** | −202 | 41 | **02** | 218 | 40 | **52** | 145 | 38 |
| **03** | 130 | 43 | **53** | −243 | 41 | **03** | −258 | 39 | **53** | −183 | 38 |
| **04** | 173 | 44 | **54** | 284 | 41 | **04** | −297 | 39 | **54** | −221 | 37 |
| **05** | −217 | 43 | **55** | 325 | 41 | **05** | 336 | 40 | **55** | 258 | 38 |
| **06** | −260 | 43 | **56** | 366 | 41 | **06** | −376 | 39 | **56** | −296 | 37 |
| **07** | −303 | 43 | **57** | 407 | 42 | **07** | −415 | 39 | **57** | 333 | 38 |
| **08** | 346 | 43 | **58** | −449 | 41 | **08** | −454 | 39 | **58** | −371 | 37 |
| **09** | 389 | 43 | **59** | −490 | 41 | **09** | 493 | 39 | **59** | 408 | 38 |
| **1010** | 432 | 43 | **1060** | −531 | 41 | **1110** | 532 | 39 | **1160** | −446 | 37 |
| **11** | 475 | 43 | **61** | −572 | 40 | **11** | 571 | 39 | **61** | 483 | 38 |
| **12** | 518 | 43 | **62** | 612 | 41 | **12** | 610 | 40 | **62** | −521 | 37 |
| **13** | −561 | 43 | **63** | 653 | 41 | **13** | −650 | 39 | **63** | −558 | 37 |
| **14** | −604 | 43 | **64** | 694 | 41 | **14** | −689 | 38 | **64** | 595 | 38 |
| **15** | −647 | 42 | **65** | −735 | 41 | **15** | 727 | 39 | **65** | −633 | 37 |
| **16** | 689 | 43 | **66** | −776 | 40 | **16** | 766 | 39 | **66** | −670 | 37 |
| **17** | 732 | 43 | **67** | 816 | 41 | **17** | 805 | 39 | **67** | 707 | 37 |
| **18** | −775 | 42 | **68** | 857 | 41 | **18** | 844 | 39 | **68** | 744 | 37 |
| **19** | 817 | 43 | **69** | −898 | 41 | **19** | 883 | 39 | **69** | 781 | 38 |
| **1020** | 860 | 43 | **1070** | 938 | 41 | **1120** | −922 | 39 | **1170** | −819 | 37 |
| **21** | −903 | 42 | **71** | −979 | 40 | **21** | −961 | 38 | **71** | −856 | 37 |
| **22** | 945 | 43 | **72** | 03019 | 41 | **22** | 999 | 39 | **72** | −893 | 37 |
| **23** | −988 | 42 | **73** | −060 | 40 | **23** | 05038 | 39 | **73** | −930 | 37 |
| **24** | 01030 | 42 | **74** | 100 | 41 | **24** | −077 | 38 | **74** | −967 | 37 |
| **25** | 072 | 43 | **75** | −141 | 40 | **25** | 115 | 39 | **75** | 07004 | 37 |
| **26** | −115 | 42 | **76** | 181 | 41 | **26** | −154 | 38 | **76** | −041 | 37 |
| **27** | 157 | 42 | **77** | −222 | 40 | **27** | 192 | 39 | **77** | −078 | 37 |
| **28** | 199 | 43 | **78** | −262 | 40 | **28** | −231 | 38 | **78** | −115 | 36 |
| **29** | −242 | 42 | **79** | 302 | 40 | **29** | 269 | 39 | **79** | 151 | 37 |
| **1030** | −284 | 42 | **1080** | 342 | 41 | **1130** | −308 | 38 | **1180** | 188 | 37 |
| **31** | −326 | 42 | **81** | −383 | 40 | **31** | 346 | 39 | **81** | −225 | 37 |
| **32** | −368 | 42 | **82** | −423 | 40 | **32** | −385 | 38 | **82** | −262 | 36 |
| **33** | 410 | 42 | **83** | −463 | 40 | **33** | −423 | 38 | **83** | 298 | 37 |
| **34** | 452 | 42 | **84** | −503 | 40 | **34** | 461 | 39 | **84** | 335 | 37 |
| **35** | 494 | 42 | **85** | −543 | 40 | **35** | −500 | 38 | **85** | −372 | 36 |
| **36** | −536 | 42 | **86** | −583 | 40 | **36** | −538 | 38 | **86** | 408 | 37 |
| **37** | −576 | 42 | **87** | −623 | 40 | **37** | 576 | 38 | **87** | 445 | 37 |
| **38** | −620 | 42 | **88** | −663 | 40 | **38** | 614 | 38 | **88** | −482 | 36 |
| **39** | −662 | 41 | **89** | −703 | 40 | **39** | 652 | 38 | **89** | 518 | 37 |
| **1040** | 703 | 42 | **1090** | −743 | 39 | **1140** | 690 | 39 | **1190** | −555 | 36 |
| **41** | 745 | 42 | **91** | 782 | 40 | **41** | −729 | 38 | **91** | 591 | 37 |
| **42** | −787 | 41 | **92** | 822 | 40 | **42** | −767 | 38 | **92** | −628 | 36 |
| **43** | 828 | 42 | **93** | 862 | 40 | **43** | −805 | 38 | **93** | 664 | 36 |
| **44** | 870 | 42 | **94** | −902 | 39 | **44** | −843 | 38 | **94** | 700 | 37 |
| **45** | −912 | 41 | **95** | 941 | 40 | **45** | −881 | 37 | **95** | −737 | 36 |
| **46** | 953 | 42 | **96** | 981 | 40 | **46** | 918 | 38 | **96** | 773 | 36 |
| **47** | −995 | 41 | **97** | 04021 | 39 | **47** | 956 | 38 | **97** | 809 | 37 |
| **48** | 02036 | 42 | **98** | 060 | 40 | **48** | 994 | 38 | **98** | −846 | 36 |
| **49** | −078 | 41 | **99** | −100 | 39 | **49** | 06032 | 38 | **99** | −882 | 36 |

**P. P.**

| | **44** | **43** |
|---|---|---|
| **1** | 4 | 4 |
| **2** | 9 | 9 |
| **3** | 13 | 13 |
| **4** | 18 | 17 |
| **5** | 22 | 22 |
| **6** | 26 | 26 |
| **7** | 31 | 30 |
| **8** | 35 | 34 |
| **9** | 40 | 39 |

| | **42** | **41** |
|---|---|---|
| **1** | 4 | 4 |
| **2** | 8 | 8 |
| **3** | 13 | 12 |
| **4** | 17 | 16 |
| **5** | 21 | 21 |
| **6** | 25 | 25 |
| **7** | 29 | 29 |
| **8** | 34 | 33 |
| **9** | 38 | 37 |

| | **40** | **39** |
|---|---|---|
| **1** | 4 | 4 |
| **2** | 8 | 8 |
| **3** | 12 | 12 |
| **4** | 16 | 16 |
| **5** | 20 | 20 |
| **6** | 24 | 23 |
| **7** | 28 | 27 |
| **8** | 32 | 31 |
| **9** | 36 | 35 |

| | **38** | **37** |
|---|---|---|
| **1** | 4 | 4 |
| **2** | 8 | 7 |
| **3** | 11 | 11 |
| **4** | 15 | 15 |
| **5** | 19 | 19 |
| **6** | 23 | 22 |
| **7** | 27 | 26 |
| **8** | 30 | 30 |
| **9** | 34 | 33 |

| | **36** |
|---|---|
| **1** | 4 |
| **2** | 7 |
| **3** | 11 |
| **4** | 14 |
| **5** | 18 |
| **6** | 22 |
| **7** | 25 |
| **8** | 29 |
| **9** | 32 |

## LOGARITHMS OF NUMBERS.—*Continued.*

| No. | Log. | Dif. | No. | Log. | Dif | No. | Log. | Dif. | No. | Log. | Dif. |
|---|---|---|---|---|---|---|---|---|---|---|---|
| **1200** | 07918 | 36 | **1250** | 09691 | 35 | **1300** | 11394 | 34 | **1350** | 13033 | 33 |
| **01** | 954 | 36 | **51** | −726 | 34 | **01** | −428 | 33 | **51** | −066 | 32 |
| **02** | 990 | 37 | **52** | 760 | 35 | **02** | 461 | 33 | **52** | −098 | 32 |
| **03** | 08027 | 36 | **53** | 795 | 35 | **03** | 494 | 34 | **53** | −130 | 32 |
| **04** | −063 | 36 | **54** | −830 | 34 | **04** | −528 | 33 | **54** | −162 | 32 |
| **05** | −099 | 36 | **55** | 864 | 35 | **05** | 561 | 33 | **55** | −194 | 32 |
| **06** | −135 | 36 | **56** | −899 | 35 | **06** | 594 | 34 | **56** | −226 | 32 |
| **07** | −171 | 36 | **57** | −934 | 34 | **07** | −628 | 33 | **57** | −258 | 32 |
| **08** | −207 | 36 | **58** | −968 | 35 | **08** | −661 | 33 | **58** | −290 | 32 |
| **09** | −243 | 36 | **59** | 10003 | 34 | **09** | −694 | 33 | **59** | −322 | 32 |
| **1210** | −279 | 35 | **1260** | 037 | 35 | **1310** | 727 | 33 | **1360** | −354 | 32 |
| **11** | 314 | 36 | **61** | −072 | 34 | **11** | 760 | 33 | **61** | −386 | 32 |
| **12** | 350 | 36 | **62** | −106 | 34 | **12** | 793 | 33 | **62** | −418 | 32 |
| **13** | 386 | 36 | **63** | 140 | 35 | **13** | 826 | 34 | **63** | −450 | 31 |
| **14** | −422 | 36 | **64** | −175 | 34 | **14** | −860 | 33 | **64** | 481 | 32 |
| **15** | −458 | 35 | **65** | 209 | 34 | **15** | −893 | 33 | **65** | 513 | 32 |
| **16** | 493 | 36 | **66** | 243 | 35 | **16** | −926 | 33 | **66** | 545 | 32 |
| **17** | 529 | 36 | **67** | −278 | 34 | **17** | −959 | 33 | **67** | −577 | 32 |
| **18** | −565 | 35 | **68** | −312 | 34 | **18** | −992 | 32 | **68** | −609 | 31 |
| **19** | 600 | 36 | **69** | 346 | 34 | **19** | 12024 | 33 | **69** | 640 | 32 |
| **1220** | −636 | 36 | **1270** | 380 | 35 | **1320** | 057 | 33 | **1370** | 672 | 32 |
| **21** | −672 | 35 | **71** | −415 | 34 | **21** | 090 | 33 | **71** | −704 | 31 |
| **22** | 707 | 36 | **72** | −449 | 34 | **22** | 123 | 33 | **72** | 735 | 32 |
| **23** | −743 | 35 | **73** | −483 | 34 | **23** | −156 | 33 | **73** | 767 | 32 |
| **24** | 778 | 36 | **74** | −517 | 34 | **24** | −189 | 33 | **74** | −799 | 31 |
| **25** | −814 | 35 | **75** | 551 | 34 | **25** | −222 | 32 | **75** | 830 | 32 |
| **26** | 849 | 35 | **76** | 585 | 34 | **26** | 254 | 33 | **76** | −862 | 31 |
| **27** | 884 | 36 | **77** | 619 | 34 | **27** | 287 | 33 | **77** | 893 | 32 |
| **28** | −920 | 35 | **78** | 653 | 34 | **28** | −320 | 32 | **78** | −925 | 31 |
| **29** | 955 | 36 | **79** | 687 | 34 | **29** | −352 | 33 | **79** | 956 | 32 |
| **1230** | −991 | 35 | **1280** | −721 | 34 | **1330** | 385 | 33 | **1380** | −988 | 31 |
| **31** | 09026 | 35 | **81** | −755 | 34 | **31** | −418 | 32 | **81** | 14019 | 32 |
| **32** | 061 | 35 | **82** | −789 | 34 | **32** | 450 | 33 | **82** | −051 | 31 |
| **33** | 096 | 36 | **83** | −823 | 34 | **33** | 483 | 33 | **83** | 082 | 32 |
| **34** | −132 | 35 | **84** | −857 | 33 | **34** | −516 | 32 | **84** | −114 | 31 |
| **35** | −167 | 35 | **85** | 890 | 34 | **35** | 548 | 33 | **85** | −145 | 31 |
| **36** | −202 | 35 | **86** | 924 | 34 | **36** | −581 | 32 | **86** | 176 | 32 |
| **37** | −237 | 35 | **87** | −958 | 34 | **37** | 613 | 33 | **87** | −208 | 31 |
| **38** | 272 | 35 | **88** | −992 | 33 | **38** | −646 | 32 | **88** | −239 | 31 |
| **39** | 307 | 35 | **89** | 11025 | 34 | **39** | 678 | 32 | **89** | 270 | 31 |
| **1240** | 342 | 35 | **1290** | −059 | 34 | **1340** | 710 | 33 | **1390** | 301 | 32 |
| **41** | 377 | 35 | **91** | −093 | 33 | **41** | −743 | 32 | **91** | 333 | 31 |
| **42** | 412 | 35 | **92** | 126 | 34 | **42** | 775 | 33 | **92** | −364 | 31 |
| **43** | 447 | 35 | **93** | −160 | 33 | **43** | −808 | 32 | **93** | 395 | 31 |
| **44** | 482 | 35 | **94** | 193 | 34 | **44** | −840 | 32 | **94** | 426 | 31 |
| **45** | −517 | 35 | **95** | −227 | 34 | **45** | 872 | 33 | **95** | 457 | 32 |
| **46** | −552 | 35 | **96** | −261 | 33 | **46** | −905 | 32 | **96** | −489 | 31 |
| **47** | −587 | 34 | **97** | −294 | 33 | **47** | −937 | 32 | **97** | −520 | 31 |
| **48** | 621 | 35 | **98** | 327 | 34 | **48** | −969 | 32 | **98** | −551 | 31 |
| **49** | 656 | 35 | **99** | −361 | 33 | **49** | 13001 | 32 | **99** | −582 | 31 |

**P. P.**

| | **36** | **35** | **34** | **33** | **32** | **31** |
|---|---|---|---|---|---|---|
| **1** | 4 | 4 | 3 | 3 | 3 | 3 |
| **2** | 7 | 7 | 7 | 7 | 6 | 6 |
| **3** | 11 | 11 | 10 | 10 | 10 | 9 |
| **4** | 14 | 14 | 14 | 13 | 13 | 12 |
| **5** | 18 | 18 | 17 | 17 | 16 | 16 |
| **6** | 22 | 21 | 20 | 20 | 19 | 19 |
| **7** | 25 | 25 | 24 | 23 | 22 | 22 |
| **8** | 29 | 28 | 27 | 26 | 26 | 25 |
| **9** | 32 | 32 | 31 | 30 | 29 | 28 |

## LOGARITHMS OF NUMBERS. — *Continued.*

| No. | Log. | Dif. | No. | Log. | Dif. | No. | Log. | Dif. | No. | Log. | Dif. |
|---|---|---|---|---|---|---|---|---|---|---|---|
| **1400** | 14613 | 31 | **1450** | 16137 | 30 | **1500** | 17609 | 29 | **1550** | 19033 | 28 |
| **01** | −644 | 31 | **51** | −167 | 30 | **01** | 638 | 29 | **51** | 061 | 28 |
| **02** | −675 | 31 | **52** | −197 | 30 | **02** | −667 | 29 | **52** | 089 | 28 |
| **03** | −706 | 31 | **53** | −227 | 29 | **03** | −696 | 29 | **53** | 117 | 28 |
| **04** | −737 | 31 | **54** | 256 | 30 | **04** | −725 | 29 | **54** | 145 | 28 |
| **05** | −768 | 31 | **55** | 286 | 30 | **05** | −754 | 28 | **55** | 173 | 28 |
| **06** | −799 | 30 | **56** | 316 | 30 | **06** | 782 | 29 | **56** | −201 | 28 |
| **07** | 829 | 31 | **57** | −346 | 30 | **07** | 811 | 29 | **57** | −229 | 28 |
| **08** | 860 | 31 | **58** | −376 | 30 | **08** | 840 | 29 | **58** | −257 | 28 |
| **09** | 191 | 31 | **59** | −406 | 29 | **09** | −869 | 29 | **59** | −285 | 27 |
| **1410** | −922 | 31 | **1460** | 435 | 30 | **1510** | −898 | 28 | **1560** | 312 | 28 |
| **11** | −953 | 30 | **61** | 465 | 30 | **11** | 926 | 29 | **61** | 340 | 28 |
| **12** | 983 | 31 | **62** | −495 | 29 | **12** | 955 | 29 | **62** | 368 | 28 |
| **13** | 15014 | 31 | **63** | 524 | 30 | **13** | −984 | 29 | **63** | −396 | 28 |
| **14** | −045 | 31 | **64** | 554 | 30 | **14** | 18013 | 28 | **64** | −424 | 27 |
| **15** | −076 | 30 | **65** | −584 | 29 | **15** | 041 | 29 | **65** | 451 | 28 |
| **16** | 106 | 31 | **66** | 613 | 30 | **16** | −070 | 29 | **66** | 479 | 28 |
| **17** | −137 | 31 | **67** | 643 | 30 | **17** | −099 | 28 | **67** | −507 | 28 |
| **18** | −168 | 30 | **68** | −673 | 29 | **18** | 127 | 29 | **68** | −535 | 27 |
| **19** | 198 | 31 | **69** | 702 | 30 | **19** | −156 | 28 | **69** | −562 | 28 |
| **1420** | −229 | 30 | **1470** | −732 | 29 | **1520** | 184 | 29 | **1570** | −590 | 28 |
| **21** | 259 | 31 | **71** | 761 | 30 | **21** | −213 | 28 | **71** | −618 | 27 |
| **22** | −290 | 30 | **72** | −791 | 29 | **22** | 241 | 29 | **72** | 645 | 28 |
| **23** | 320 | 31 | **73** | 820 | 30 | **23** | −270 | 28 | **73** | −673 | 27 |
| **24** | −351 | 30 | **74** | −850 | 29 | **24** | 298 | 29 | **74** | 700 | 28 |
| **25** | 381 | 31 | **75** | 879 | 30 | **25** | −327 | 28 | **75** | 728 | 28 |
| **26** | −412 | 30 | **76** | −909 | 29 | **26** | 355 | 29 | **76** | −756 | 27 |
| **27** | 442 | 31 | **77** | 938 | 29 | **27** | −384 | 28 | **77** | 783 | 28 |
| **28** | −473 | 30 | **78** | 967 | 30 | **28** | 412 | 29 | **78** | −811 | 27 |
| **29** | 503 | 31 | **79** | −997 | 29 | **29** | −441 | 28 | **79** | 838 | 28 |
| **1430** | −534 | 30 | **1480** | 17026 | 30 | **1530** | 469 | 29 | **1580** | −866 | 27 |
| **31** | −564 | 30 | **81** | −056 | 29 | **31** | −498 | 28 | **81** | 893 | 28 |
| **32** | 594 | 31 | **82** | −085 | 29 | **32** | −526 | 28 | **82** | −921 | 27 |
| **33** | 625 | 30 | **83** | 114 | 29 | **33** | 554 | 29 | **83** | 948 | 28 |
| **34** | −655 | 30 | **84** | 143 | 30 | **34** | −583 | 28 | **84** | −976 | 27 |
| **35** | 685 | 30 | **85** | −173 | 29 | **35** | −611 | 28 | **85** | 20003 | 27 |
| **36** | 715 | 31 | **86** | −202 | 29 | **36** | 639 | 28 | **86** | 030 | 28 |
| **37** | −746 | 30 | **87** | 231 | 29 | **37** | 667 | 29 | **87** | −058 | 27 |
| **38** | −776 | 30 | **88** | 260 | 29 | **38** | −696 | 28 | **88** | 085 | 27 |
| **39** | 806 | 30 | **89** | 289 | 30 | **39** | −724 | 28 | **89** | 112 | 28 |
| **1440** | 836 | 30 | **1490** | −319 | 29 | **1540** | 752 | 28 | **1590** | −140 | 27 |
| **41** | 866 | 31 | **91** | −348 | 29 | **41** | 780 | 28 | **91** | 167 | 27 |
| **42** | −897 | 30 | **92** | −377 | 29 | **42** | 808 | 29 | **92** | 194 | 28 |
| **43** | −927 | 30 | **93** | −406 | 29 | **43** | −837 | 28 | **93** | −222 | 27 |
| **44** | −957 | 30 | **94** | 435 | 29 | **44** | −865 | 28 | **94** | −249 | 27 |
| **45** | −987 | 30 | **95** | 464 | 29 | **45** | −893 | 28 | **95** | 276 | 27 |
| **46** | 16017 | 30 | **96** | 493 | 29 | **46** | −921 | 28 | **96** | 303 | 27 |
| **47** | −047 | 30 | **97** | 522 | 29 | **47** | 949 | 28 | **97** | 330 | 28 |
| **48** | −077 | 30 | **98** | 551 | 29 | **48** | 977 | 28 | **98** | −358 | 27 |
| **49** | −107 | 31 | **99** | 580 | 29 | **49** | 19005 | 28 | **99** | −385 | 27 |

P. P.

| | **31** | **30** | **29** | **28** | **27** |
|---|---|---|---|---|---|
| **1** | 3 | 3 | 3 | 3 | 3 |
| **2** | 6 | 6 | 6 | 6 | 5 |
| **3** | 9 | 9 | 9 | 8 | 8 |
| **4** | 12 | 12 | 12 | 11 | 11 |
| **5** | 16 | 15 | 15 | 14 | 14 |
| **6** | 19 | 18 | 17 | 17 | 16 |
| **7** | 22 | 21 | 20 | 20 | 19 |
| **8** | 25 | 24 | 23 | 22 | 22 |
| **9** | 28 | 27 | 26 | 25 | 24 |

## LOGARITHMS OF NUMBERS.—*Continued.*

| No. | Log. | Dif. | No. | Log. | Dif. | No. | Log. | Dif. | No. | Log. | Dif. |
|---|---|---|---|---|---|---|---|---|---|---|---|
| 1600 | 20412 | 27 | 1650 | 21748 | 27 | 1700 | 23045 | 25 | 1750 | 24304 | 49 |
| 01 | 439 | 27 | 51 | −775 | 26 | 01 | 070 | 26 | 52 | 353 | 50 |
| 02 | 466 | 27 | 52 | 801 | 26 | 02 | −096 | 25 | 54 | −403 | 49 |
| 03 | 493 | 27 | 53 | 827 | 27 | 03 | 121 | 26 | 56 | 452 | 50 |
| 04 | 520 | 28 | 54 | −854 | 26 | 04 | −147 | 25 | 58 | −502 | 49 |
| 05 | −548 | 27 | 55 | −880 | 26 | 05 | 172 | 26 | 1760 | 551 | 50 |
| 06 | −575 | 27 | 56 | 906 | 26 | 06 | −198 | 25 | 62 | −601 | 49 |
| 07 | −602 | 27 | 57 | 932 | 26 | 07 | 223 | 26 | 64 | −650 | 49 |
| 08 | −629 | 27 | 58 | 958 | 27 | 08 | −249 | 25 | 66 | 699 | 49 |
| 09 | −656 | 27 | 59 | −985 | 26 | 09 | 274 | 26 | 68 | 748 | 49 |
| 1610 | −683 | 27 | 1660 | 22011 | 26 | 1710 | −300 | 25 | 1770 | 797 | 49 |
| 11 | −710 | 27 | 61 | −037 | 26 | 11 | 325 | 25 | 72 | 846 | 49 |
| 12 | −737 | 26 | 62 | 063 | 26 | 12 | 350 | 26 | 74 | 895 | 49 |
| 13 | 763 | 27 | 63 | 089 | 26 | 13 | −376 | 25 | 76 | 944 | 49 |
| 14 | 790 | 27 | 64 | 115 | 26 | 14 | 401 | 25 | 78 | 993 | 49 |
| 15 | 817 | 27 | 65 | 141 | 26 | 15 | 426 | 26 | 1780 | 25042 | 49 |
| 16 | 844 | 27 | 66 | 167 | 27 | 16 | −452 | 25 | 82 | −091 | 48 |
| 17 | 871 | 27 | 67 | −194 | 26 | 17 | 477 | 25 | 84 | 139 | 49 |
| 18 | −898 | 27 | 68 | −220 | 26 | 18 | 502 | 26 | 86 | 188 | 49 |
| 19 | −925 | 27 | 69 | −246 | 26 | 19 | −528 | 25 | 88 | 237 | 48 |
| 1620 | −952 | 26 | 1670 | −272 | 26 | 1720 | −553 | 25 | 1790 | 285 | 49 |
| 21 | 978 | 27 | 71 | −298 | 26 | 21 | 578 | 25 | 92 | −334 | 48 |
| 22 | 21005 | 27 | 72 | −324 | 26 | 22 | 603 | 26 | 94 | 382 | 49 |
| 23 | −032 | 27 | 73 | −350 | 26 | 23 | −629 | 25 | 96 | −431 | 48 |
| 24 | −059 | 26 | 74 | −376 | 25 | 24 | −654 | 25 | 98 | −479 | 48 |
| 25 | 085 | 27 | 75 | 401 | 26 | 25 | −679 | 25 | 1800 | 527 | 48 |
| 26 | 112 | 27 | 76 | 427 | 26 | 26 | 704 | 25 | 02 | 575 | 49 |
| 27 | −139 | 26 | 77 | 453 | 26 | 27 | 729 | 25 | 04 | −624 | 48 |
| 28 | 165 | 27 | 78 | 479 | 26 | 28 | 754 | 25 | 06 | −672 | 48 |
| 29 | 192 | 27 | 79 | 505 | 26 | 29 | 779 | 26 | 08 | −720 | 48 |
| 1630 | −219 | 26 | 1680 | −531 | 26 | 1730 | −805 | 25 | 1810 | −768 | 48 |
| 31 | 245 | 27 | 81 | −557 | 26 | 31 | −830 | 25 | 12 | −816 | 48 |
| 32 | 272 | 27 | 82 | −583 | 25 | 32 | −855 | 25 | 14 | −864 | 48 |
| 33 | −299 | 26 | 83 | 608 | 26 | 33 | −880 | 25 | 16 | −912 | 47 |
| 34 | 325 | 27 | 84 | 634 | 26 | 34 | −905 | 25 | 18 | 959 | 48 |
| 35 | −352 | 26 | 85 | −660 | 26 | 35 | −930 | 25 | 1820 | 26007 | 48 |
| 36 | 378 | 27 | 86 | −686 | 26 | 36 | −955 | 25 | 22 | −055 | 47 |
| 37 | −405 | 26 | 87 | −712 | 25 | 37 | −980 | 25 | 24 | 102 | 48 |
| 38 | 431 | 27 | 88 | 737 | 26 | 38 | 24005 | 25 | 26 | 150 | 48 |
| 39 | −458 | 26 | 89 | −763 | 26 | 39 | −030 | 25 | 28 | −198 | 47 |
| 1640 | 484 | 27 | 1690 | 789 | 25 | 1740 | −055 | 25 | 1830 | 245 | 48 |
| 41 | −511 | 26 | 91 | 814 | 26 | 41 | −080 | 25 | 32 | −293 | 47 |
| 42 | 537 | 27 | 92 | 840 | 26 | 42 | −105 | 25 | 34 | −340 | 47 |
| 43 | −564 | 26 | 93 | −866 | 25 | 43 | −130 | 25 | 36 | 387 | 48 |
| 44 | 590 | 27 | 94 | 891 | 26 | 44 | −155 | 25 | 38 | −435 | 47 |
| 45 | −617 | 26 | 95 | −917 | 26 | 45 | −180 | 24 | 1840 | −482 | 47 |
| 46 | −643 | 26 | 96 | −943 | 25 | 46 | 204 | 25 | 42 | −529 | 47 |
| 47 | 669 | 27 | 97 | 968 | 26 | 47 | 229 | 25 | 44 | 576 | 47 |
| 48 | −696 | 26 | 98 | −994 | 25 | 48 | 254 | 25 | 46 | 623 | 47 |
| 49 | 722 | 26 | 99 | 23019 | 26 | 49 | −279 | 25 | 48 | 670 | 47 |

P. P.

| | 50 | 49 |
|---|---|---|
| 1 | 3 | 2 |
| 2 | 5 | 5 |
| 3 | 8 | 7 |
| 4 | 10 | 10 |
| 5 | 13 | 12 |
| 6 | 15 | 15 |
| 7 | 18 | 17 |
| 8 | 20 | 20 |
| 9 | 23 | 22 |
| 10 | 25 | 25 |

| | 48 | 47 |
|---|---|---|
| 1 | 2 | 2 |
| 2 | 5 | 5 |
| 3 | 7 | 7 |
| 4 | 10 | 9 |
| 5 | 12 | 12 |
| 6 | 14 | 14 |
| 7 | 17 | 16 |
| 8 | 19 | 19 |
| 9 | 22 | 21 |
| 10 | 24 | 24 |

| | 27 | 26 |
|---|---|---|
| 1 | 3 | 3 |
| 2 | 5 | 5 |
| 3 | 8 | 8 |
| 4 | 11 | 10 |
| 5 | 14 | 13 |
| 6 | 16 | 16 |
| 7 | 19 | 18 |
| 8 | 22 | 21 |
| 9 | 24 | 23 |

| | 25 |
|---|---|
| 1 | 3 |
| 2 | 5 |
| 3 | 8 |
| 4 | 10 |
| 5 | 13 |
| 6 | 15 |
| 7 | 18 |
| 8 | 20 |
| 9 | 23 |

| | 24 |
|---|---|
| 1 | 2 |
| 2 | 5 |
| 3 | 7 |
| 4 | 10 |
| 5 | 12 |
| 6 | 14 |
| 7 | 17 |
| 8 | 19 |
| 9 | 22 |

## LOGARITHMS OF NUMBERS.—*Continued.*

| No. | Log. | Dif. | No. | Log. | Dif. | No. | Log. | Dif. | No. | Log. | Dif. |
|---|---|---|---|---|---|---|---|---|---|---|---|
| **1850** | 26717 | 47 | **1950** | 29003 | 45 | **2050** | 31175 | 43 | **2150** | 33244 | 40 |
| **52** | 764 | 47 | **52** | −048 | 44 | **52** | −218 | 42 | **52** | 284 | 41 |
| **54** | −811 | 47 | **54** | 092 | 45 | **54** | 260 | 42 | **54** | −325 | 40 |
| **56** | −858 | 47 | **56** | −137 | 44 | **56** | 302 | 43 | **56** | −365 | 40 |
| **58** | −905 | 46 | **58** | 181 | 45 | **58** | −345 | 42 | **58** | 405 | 40 |
| **1860** | 951 | 47 | **1960** | −226 | 44 | **2060** | −387 | 42 | **2160** | 445 | 41 |
| **62** | −998 | 47 | **62** | −270 | 44 | **62** | −429 | 42 | **62** | −486 | 40 |
| **64** | 27045 | 46 | **64** | 314 | 44 | **64** | −471 | 42 | **64** | −526 | 40 |
| **66** | 091 | 47 | **66** | 358 | 45 | **66** | 513 | 42 | **66** | −566 | 40 |
| **68** | −138 | 46 | **68** | −403 | 44 | **68** | 555 | 42 | **68** | −606 | 40 |
| **1870** | 184 | 47 | **1970** | −447 | 44 | **2070** | 597 | 42 | **2170** | −646 | 40 |
| **72** | −231 | 46 | **72** | −491 | 44 | **72** | −639 | 42 | **72** | −686 | 40 |
| **74** | −277 | 46 | **74** | −535 | 44 | **74** | −681 | 42 | **74** | −726 | 40 |
| **76** | 323 | 47 | **76** | −579 | 44 | **76** | −723 | 42 | **76** | −766 | 40 |
| **78** | −370 | 46 | **78** | −623 | 44 | **78** | −765 | 41 | **78** | −806 | 40 |
| **1880** | −416 | 46 | **1980** | −667 | 43 | **2080** | 806 | 42 | **2180** | −846 | 39 |
| **82** | −462 | 46 | **82** | 710 | 44 | **82** | 848 | 42 | **82** | 885 | 40 |
| **84** | 508 | 46 | **84** | 754 | 44 | **84** | −890 | 41 | **84** | 925 | 40 |
| **86** | 554 | 46 | **86** | −798 | 44 | **86** | 932 | 42 | **86** | 965 | 40 |
| **88** | 600 | 46 | **88** | −842 | 43 | **88** | 973 | 42 | **88** | 34005 | 39 |
| **1890** | 646 | 46 | **1990** | 885 | 44 | **2090** | 32015 | 41 | **2190** | 044 | 40 |
| **92** | 692 | 46 | **92** | −929 | 44 | **92** | 056 | 42 | **92** | 084 | 40 |
| **94** | −738 | 46 | **94** | −973 | 43 | **94** | −098 | 41 | **94** | −124 | 39 |
| **96** | −784 | 46 | **96** | 30016 | 44 | **96** | 139 | 42 | **96** | 163 | 40 |
| **98** | −830 | 45 | **98** | −060 | 43 | **98** | −181 | 41 | **98** | −203 | 39 |
| **1900** | 875 | 46 | **2000** | −103 | 43 | **2100** | −222 | 41 | **2200** | 242 | 40 |
| **02** | 921 | 46 | **02** | 146 | 44 | **02** | 263 | 42 | **02** | −282 | 39 |
| **04** | −967 | 45 | **04** | −190 | 43 | **04** | −305 | 41 | **04** | 321 | 40 |
| **06** | 28012 | 46 | **06** | 233 | 43 | **06** | −346 | 41 | **06** | −361 | 39 |
| **08** | −058 | 45 | **08** | 276 | 44 | **08** | 387 | 41 | **08** | −400 | 39 |
| **1910** | 103 | 46 | **2010** | −320 | 43 | **2110** | 428 | 41 | **2210** | 439 | 40 |
| **12** | −149 | 45 | **12** | −363 | 43 | **12** | 469 | 41 | **12** | −479 | 39 |
| **14** | 194 | 46 | **14** | −406 | 43 | **14** | 510 | 42 | **14** | −518 | 39 |
| **16** | −240 | 45 | **16** | 449 | 43 | **16** | −552 | 41 | **16** | −557 | 39 |
| **18** | −285 | 45 | **18** | 292 | 43 | **18** | −593 | 41 | **18** | 596 | 39 |
| **1920** | −330 | 45 | **2020** | 535 | 43 | **2120** | −634 | 41 | **2220** | 635 | 39 |
| **22** | 375 | 46 | **22** | 578 | 43 | **22** | −675 | 40 | **22** | 674 | 39 |
| **24** | −421 | 45 | **24** | 621 | 43 | **24** | 715 | 41 | **24** | 713 | 40 |
| **26** | −466 | 45 | **26** | −664 | 43 | **26** | 756 | 41 | **26** | −753 | 39 |
| **28** | −511 | 45 | **28** | −707 | 43 | **28** | 797 | 41 | **28** | −792 | 38 |
| **1930** | −556 | 45 | **2030** | −750 | 42 | **2130** | −838 | 41 | **2230** | 830 | 39 |
| **32** | −601 | 45 | **32** | 792 | 42 | **32** | −879 | 40 | **32** | 869 | 39 |
| **34** | −646 | 45 | **34** | 835 | 43 | **34** | 919 | 41 | **34** | 908 | 39 |
| **36** | −691 | 44 | **36** | −878 | 43 | **36** | 960 | 41 | **36** | 947 | 39 |
| **38** | 735 | 45 | **38** | 920 | 42 | **38** | 33001 | 40 | **38** | 986 | 39 |
| **1940** | 780 | 45 | **2040** | 963 | 43 | **2140** | 041 | 41 | **2240** | 35025 | 39 |
| **42** | −825 | 45 | **42** | 31006 | 42 | **42** | −082 | 40 | **42** | −064 | 38 |
| **44** | −870 | 44 | **44** | 048 | 43 | **44** | 122 | 41 | **44** | 102 | 39 |
| **46** | 941 | 45 | **46** | −091 | 42 | **46** | −163 | 40 | **46** | −141 | 39 |
| **48** | −959 | 44 | **48** | −133 | 42 | **48** | 203 | 41 | **48** | −180 | 38 |

**P. P.**

| | **47** | **46** |
|---|---|---|
| **1** | 2 | 2 |
| **2** | 5 | 5 |
| **3** | 7 | 7 |
| **4** | 9 | 9 |
| **5** | 12 | 12 |
| **6** | 14 | 14 |
| **7** | 16 | 16 |
| **8** | 19 | 18 |
| **9** | 21 | 21 |
| **10** | 24 | 23 |

| | **45** | **44** |
|---|---|---|
| **1** | 2 | 2 |
| **2** | 5 | 5 |
| **3** | 7 | 7 |
| **4** | 9 | 9 |
| **5** | 11 | 11 |
| **6** | 14 | 13 |
| **7** | 16 | 15 |
| **8** | 18 | 18 |
| **9** | 20 | 20 |
| **10** | 23 | 22 |

| | **43** | **42** |
|---|---|---|
| **1** | 2 | 2 |
| **2** | 4 | 4 |
| **3** | 6 | 6 |
| **4** | 9 | 8 |
| **5** | 11 | 11 |
| **6** | 13 | 13 |
| **7** | 15 | 15 |
| **8** | 17 | 17 |
| **9** | 19 | 19 |
| **10** | 22 | 21 |

| | **41** | **40** |
|---|---|---|
| **1** | 2 | 2 |
| **2** | 4 | 4 |
| **3** | 6 | 6 |
| **4** | 8 | 8 |
| **5** | 10 | 10 |
| **6** | 12 | 12 |
| **7** | 14 | 14 |
| **8** | 16 | 16 |
| **9** | 18 | 18 |
| **10** | 21 | 20 |

| | **39** | **38** |
|---|---|---|
| **1** | 2 | 2 |
| **2** | 4 | 4 |
| **3** | 6 | 6 |
| **4** | 8 | 8 |
| **5** | 10 | 10 |
| **6** | 12 | 11 |
| **7** | 14 | 13 |
| **8** | 16 | 15 |
| **9** | 18 | 16 |
| **10** | 20 | 17 |

## LOGARITHMS OF NUMBERS.—*Continued.*

| No. | Log. | Dif. | No. | Log. | Dif. | No. | Log. | Dif. | No. | Log. | Dif. |
|---|---|---|---|---|---|---|---|---|---|---|---|
| **2250** | 33218 | 39 | **2350** | 37107 | 37 | **2450** | 38917 | 35 | **2550** | 40654 | 34 |
| **52** | −257 | 38 | **52** | −144 | 37 | **52** | 952 | 35 | **52** | 688 | 34 |
| **54** | 295 | 39 | **54** | −181 | 37 | **54** | 987 | 36 | **54** | 722 | 34 |
| **56** | −334 | 38 | **56** | −218 | 36 | **56** | 39023 | 35 | **56** | 756 | 34 |
| **58** | 372 | 39 | **58** | 254 | 37 | **58** | 058 | 36 | **58** | 790 | 34 |
| **2260** | −411 | 38 | **2360** | 291 | 37 | **2460** | −094 | 35 | **2560** | −824 | 34 |
| **62** | 449 | 39 | **62** | −328 | 37 | **62** | −129 | 35 | **62** | −858 | 34 |
| **64** | −488 | 38 | **64** | −365 | 36 | **64** | 164 | 35 | **64** | −892 | 34 |
| **66** | −526 | 38 | **66** | 401 | 37 | **66** | 199 | 35 | **66** | −926 | 34 |
| **68** | 564 | 39 | **68** | 438 | 37 | **68** | −235 | 35 | **68** | −960 | 33 |
| **2270** | −603 | 38 | **2370** | −475 | 36 | **2470** | −270 | 35 | **2570** | 993 | 34 |
| **72** | −641 | 38 | **72** | 511 | 37 | **72** | −305 | 35 | **72** | 41027 | 34 |
| **74** | 679 | 38 | **74** | 548 | 37 | **74** | −340 | 35 | **74** | −061 | 34 |
| **76** | 717 | 38 | **76** | −585 | 36 | **76** | 375 | 35 | **76** | −095 | 33 |
| **78** | 755 | 38 | **78** | 621 | 37 | **78** | 410 | 35 | **78** | 128 | 34 |
| **2280** | 793 | 39 | **2380** | −658 | 36 | **2480** | 455 | 35 | **2580** | −162 | 34 |
| **82** | −832 | 38 | **82** | 694 | 37 | **82** | 480 | 35 | **82** | −196 | 33 |
| **84** | −870 | 38 | **84** | −731 | 36 | **84** | 515 | 35 | **84** | 229 | 34 |
| **86** | −908 | 38 | **86** | 767 | 36 | **86** | 550 | 35 | **86** | −263 | 33 |
| **88** | −946 | 38 | **88** | 803 | 37 | **88** | 585 | 35 | **88** | 296 | 34 |
| **2290** | −984 | 37 | **2390** | −840 | 36 | **2490** | −620 | 35 | **2590** | −330 | 33 |
| **92** | 36021 | 38 | **92** | 876 | 36 | **92** | −655 | 35 | **92** | 363 | 34 |
| **94** | 059 | 38 | **94** | 912 | 37 | **94** | −690 | 34 | **94** | −397 | 33 |
| **96** | 097 | 38 | **96** | −949 | 36 | **96** | 724 | 35 | **96** | 430 | 34 |
| **98** | 135 | 38 | **98** | −985 | 36 | **98** | 759 | 35 | **98** | −464 | 33 |
| **2300** | −173 | 38 | **2400** | 38021 | 36 | **2500** | 794 | 35 | **2600** | 497 | 34 |
| **02** | −211 | 37 | **02** | 057 | 36 | **02** | −829 | 34 | **02** | −531 | 33 |
| **04** | 248 | 38 | **04** | 093 | 37 | **04** | 863 | 35 | **04** | 564 | 33 |
| **06** | −286 | 38 | **06** | −130 | 36 | **06** | 898 | 35 | **06** | 597 | 34 |
| **08** | −324 | 37 | **08** | −165 | 36 | **08** | −933 | 34 | **08** | −631 | 33 |
| **2310** | 361 | 38 | **2410** | −202 | 36 | **2510** | 967 | 35 | **2610** | 664 | 33 |
| **12** | −399 | 37 | **12** | −238 | 36 | **12** | 40002 | 35 | **12** | 697 | 34 |
| **14** | 436 | 38 | **14** | −274 | 36 | **14** | −037 | 34 | **14** | −731 | 33 |
| **16** | −474 | 37 | **16** | −310 | 36 | **16** | 071 | 35 | **16** | −764 | 33 |
| **18** | 511 | 38 | **18** | −346 | 36 | **18** | −106 | 34 | **18** | −797 | 33 |
| **2320** | −549 | 37 | **2420** | −382 | 35 | **2520** | 140 | 35 | **2620** | 830 | 33 |
| **22** | 586 | 38 | **22** | 417 | 36 | **22** | −175 | 34 | **22** | 863 | 33 |
| **24** | −624 | 37 | **24** | 453 | 36 | **24** | −209 | 34 | **24** | 896 | 33 |
| **26** | −661 | 37 | **26** | 489 | 36 | **26** | 243 | 35 | **26** | 929 | 34 |
| **28** | 698 | 38 | **28** | −525 | 36 | **28** | −278 | 34 | **28** | −963 | 33 |
| **2330** | −736 | 37 | **2430** | −561 | 35 | **2530** | 312 | 34 | **2630** | −996 | 33 |
| **32** | −773 | 37 | **32** | 596 | 36 | **32** | 346 | 35 | **32** | 42029 | 33 |
| **34** | 810 | 37 | **34** | 632 | 36 | **34** | −381 | 34 | **34** | −062 | 33 |
| **36** | 847 | 37 | **36** | −668 | 35 | **36** | −415 | 34 | **36** | −095 | 32 |
| **38** | 884 | 38 | **38** | 703 | 36 | **38** | 449 | 34 | **38** | 127 | 33 |
| **2340** | −922 | 37 | **2440** | −739 | 36 | **2540** | 483 | 35 | **2640** | 160 | 33 |
| **42** | −959 | 37 | **42** | −775 | 35 | **42** | −518 | 34 | **42** | 193 | 33 |
| **44** | −996 | 37 | **44** | 810 | 36 | **44** | −552 | 34 | **44** | 226 | 33 |
| **46** | 37033 | 37 | **46** | −846 | 35 | **46** | −586 | 34 | **46** | −259 | 33 |
| **48** | −070 | 37 | **48** | 881 | 36 | **48** | −620 | 34 | **48** | −292 | 33 |

**P. P.**

| | **39** | **38** |
|---|---|---|
| **1** | 2 | 2 |
| **2** | 4 | 4 |
| **3** | 6 | 6 |
| **4** | 8 | 8 |
| **5** | 10 | 10 |
| **6** | 12 | 11 |
| **7** | 14 | 13 |
| **8** | 16 | 15 |
| **9** | 18 | 17 |
| **10** | 20 | 19 |

| | **37** | **36** |
|---|---|---|
| **1** | 2 | 2 |
| **2** | 4 | 4 |
| **3** | 6 | 5 |
| **4** | 7 | 7 |
| **5** | 9 | 9 |
| **6** | 11 | 11 |
| **7** | 13 | 13 |
| **8** | 15 | 14 |
| **9** | 17 | 16 |
| **10** | 19 | 18 |

| | **35** | **34** |
|---|---|---|
| **1** | 2 | 2 |
| **2** | 4 | 3 |
| **3** | 5 | 5 |
| **4** | 7 | 7 |
| **5** | 9 | 9 |
| **6** | 11 | 10 |
| **7** | 12 | 12 |
| **8** | 14 | 14 |
| **9** | 16 | 15 |
| **10** | 18 | 17 |

| | **33** |
|---|---|
| **1** | 2 |
| **2** | 3 |
| **3** | 5 |
| **4** | 7 |
| **5** | 8 |
| **6** | 10 |
| **7** | 12 |
| **8** | 13 |
| **9** | 15 |
| **10** | 17 |

| | **32** |
|---|---|
| **1** | 2 |
| **2** | 3 |
| **3** | 5 |
| **4** | 6 |
| **5** | 8 |
| **6** | 10 |
| **7** | 11 |
| **8** | 13 |
| **9** | 14 |
| **10** | 16 |

## LOGARITHMS OF NUMBERS.—*Continued.*

| No. | Log. | Dif. | No. | Log. | Dif. | No. | Log. | Dif. | No. | Log. | Dif. |
|---|---|---|---|---|---|---|---|---|---|---|---|
| **2650** | 42325 | 32 | **2750** | 43933 | 32 | **2850** | 45484 | 31 | **2950** | 46982 | 30 |
| **52** | 357 | 33 | **52** | −965 | 31 | **52** | −515 | 30 | **52** | 47012 | 29 |
| **54** | 390 | 33 | **54** | 996 | 32 | **54** | 545 | 31 | **54** | 041 | 29 |
| **56** | −423 | ·32 | **56** | 44028 | 31 | **56** | −576 | 30 | **56** | 070 | 30 |
| **58** | 455 | 33 | **58** | 059 | 32 | **58** | 606 | 31 | **58** | −100 | 29 |
| **2660** | 488 | 33 | **2760** | −091 | 31 | **2860** | −637 | 30 | **2960** | 129 | 30 |
| **62** | −521 | 32 | **62** | 122 | 32 | **62** | −667 | 30 | **62** | −159 | 29 |
| **64** | 553 | 33 | **64** | −154 | 31 | **64** | 697 | 31 | **64** | −188 | 29 |
| **66** | 586 | 33 | **66** | 185 | 32 | **66** | −728 | 30 | **66** | 217 | 29 |
| **68** | −619 | 32 | **68** | −217 | 31 | **68** | −758 | 30 | **68** | 246 | 30 |
| **2670** | 651 | 33 | **2770** | −248 | 31 | **2870** | 788 | 30 | **2970** | −276 | 29 |
| **72** | −684 | 32 | **72** | 279 | 32 | **72** | 818 | 31 | **72** | −305 | 29 |
| **74** | 716 | 33 | **74** | −311 | 31 | **74** | −849 | 30 | **74** | 334 | 29 |
| **76** | −749 | 32 | **76** | −342 | 31 | **76** | −879 | 30 | **76** | 363 | 29 |
| **78** | 781 | 32 | **78** | 373 | 31 | **78** | 909 | 30 | **78** | 392 | 30 |
| **2680** | 813 | 33 | **2780** | 404 | 32 | **2880** | 939 | 30 | **2980** | −422 | 29 |
| **82** | −846 | 32 | **82** | −436 | 31 | **82** | 969 | 31 | **82** | −451 | 29 |
| **84** | 87 | 33 | **84** | −467 | 31 | **84** | 46000 | 30 | **84** | −480 | 29 |
| **86** | −911 | 32 | **86** | 498 | 31 | **86** | −030 | 30 | **86** | −509 | 29 |
| **88** | −943 | 32 | **88** | 529 | 31 | **88** | −060 | 30 | **88** | 538 | 29 |
| **2690** | 975 | 33 | **2790** | 560 | 32 | **2890** | −090 | 30 | **2990** | 567 | 29 |
| **92** | 43008 | 32 | **92** | −592 | 31 | **92** | −120 | 30 | **92** | 596 | 29 |
| **94** | −040 | 32 | **94** | −623 | 31 | **94** | −150 | 30 | **94** | 625 | 29 |
| **96** | −072 | 32 | **96** | −654 | 31 | **96** | −180 | 30 | **96** | 654 | 29 |
| **98** | 104 | 32 | **98** | −685 | 31 | **98** | −210 | 30 | **98** | 683 | 29 |
| **2700** | 136 | 33 | **2800** | −716 | 31 | **2900** | −240 | 30 | **3000** | 712 | 29 |
| **02** | −169 | 32 | **02** | −747 | 31 | **02** | −270 | 30 | **02** | 741 | 29 |
| **04** | −201 | 32 | **04** | −778 | 31 | **04** | −300 | 30 | **04** | −770 | 29 |
| **06** | −233 | 32 | **06** | −809 | 31 | **06** | −330 | 29 | **06** | −799 | 29 |
| **08** | −265 | 32 | **08** | −840 | 31 | **08** | 359 | 30 | **08** | −828 | 29 |
| **2710** | −297 | 32 | **2810** | −871 | 31 | **2910** | 389 | 30 | **3010** | −857 | 28 |
| **12** | −329 | 32 | **12** | −902 | 30 | **12** | 419 | 30 | **12** | 885 | 29 |
| **14** | −361 | 32 | **14** | 932 | 31 | **14** | −449 | 30 | **14** | 914 | 29 |
| **16** | −393 | 32 | **16** | 963 | 31 | **16** | −479 | 30 | **16** | 943 | 29 |
| **18** | −425 | 32 | **18** | 994 | 31 | **18** | −509 | 29 | **18** | −972 | 29 |
| **2720** | −457 | 32 | **2820** | 45025 | 31 | **2920** | 538 | 30 | **3020** | 48001 | 28 |
| **22** | −489 | 32 | **22** | −056 | 30 | **22** | 568 | 30 | **22** | 029 | 29 |
| **24** | −521 | 32 | **24** | 086 | 31 | **24** | −598 | 29 | **24** | 058 | 29 |
| **26** | −553 | 31 | **26** | 117 | 31 | **26** | 627 | 30 | **26** | −087 | 29 |
| **28** | 584 | 32 | **28** | −148 | 31 | **28** | 657 | 30 | **28** | −116 | 28 |
| **2730** | 616 | 32 | **2830** | −179 | 30 | **2930** | −687 | 29 | **3030** | 144 | 29 |
| **32** | 648 | 32 | **32** | 209 | 31 | **32** | 716 | 30 | **32** | −173 | 29 |
| **34** | −680 | 32 | **34** | −240 | 31 | **34** | 746 | 30 | **34** | −202 | 28 |
| **36** | −712 | 31 | **36** | −271 | 30 | **36** | −776 | 29 | **36** | 230 | 29 |
| **38** | 743 | 32 | **38** | 301 | 31 | **38** | 805 | 30 | **38** | −259 | 28 |
| **2740** | 775 | 32 | **2840** | −332 | 30 | **2940** | −835 | 29 | **3040** | 287 | 29 |
| **42** | −807 | 31 | **42** | 362 | 31 | **42** | 864 | 30 | **42** | −316 | 28 |
| **44** | 838 | 32 | **44** | −393 | 30 | **44** | −894 | 29 | **44** | 344 | 29 |
| **46** | 870 | 32 | **46** | 423 | 31 | **46** | 923 | 30 | **46** | −373 | 28 |
| **48** | −902 | 31 | **48** | −454 | 30 | **48** | −953 | 29 | **48** | 401 | 29 |

P. P.

| | **33** | **32** |
|---|---|---|
| **1** | 2 | 2 |
| **2** | 3 | 3 |
| **3** | 5 | 5 |
| **4** | 7 | 6 |
| **5** | 8 | 8 |
| **6** | 10 | 10 |
| **7** | 12 | 11 |
| **8** | 13 | 13 |
| **9** | 15 | 14 |
| **10** | 17 | 16 |

| | **31** |
|---|---|
| **1** | 2 |
| **2** | 3 |
| **3** | 5 |
| **4** | 6 |
| **5** | 8 |
| **6** | 9 |
| **7** | 11 |
| **8** | 12 |
| **9** | 14 |
| **10** | 16 |

| | **30** |
|---|---|
| **1** | 2 |
| **2** | 3 |
| **3** | 5 |
| **4** | 6 |
| **5** | 8 |
| **6** | 9 |
| **7** | 11 |
| **8** | 12 |
| **9** | 14 |
| **10** | 15 |

| | **29** |
|---|---|
| **1** | 1 |
| **2** | 3 |
| **3** | 4 |
| **4** | 6 |
| **5** | 7 |
| **6** | 9 |
| **7** | 10 |
| **8** | 12 |
| **9** | 13 |
| **10** | 15 |

| | **28** |
|---|---|
| **1** | 1 |
| **2** | 3 |
| **3** | 4 |
| **4** | 6 |
| **5** | 7 |
| **6** | 8 |
| **7** | 10 |
| **8** | 11 |
| **9** | 13 |
| **10** | 14 |

## LOGARITHMS OF NUMBERS.—*Continued.*

| No. | Log. | Dif. | No. | Log. | Dif. | No. | Log. | Dif. | No. | Log. | Dif. |
|---|---|---|---|---|---|---|---|---|---|---|---|
| **3050** | 48430 | 28 | **3150** | 49831 | 28 | **3250** | 51188 | 27 | **3350** | 52504 | 26 |
| **52** | 458 | 29 | **52** | −859 | 27 | **52** | 215 | 27 | **52** | 530 | 26 |
| **54** | −487 | 28 | **54** | 886 | 28 | **54** | −242 | 26 | **54** | 556 | 26 |
| **56** | 515 | 29 | **56** | −914 | 27 | **56** | 268 | 27 | **56** | 582 | 26 |
| **58** | −544 | 28 | **58** | 941 | 28 | **58** | 295 | 27 | **58** | 608 | 26 |
| **3060** | 572 | 29 | **3160** | −969 | 27 | **3260** | −322 | 26 | **3360** | −634 | 26 |
| **62** | −601 | 28 | **62** | 996 | 28 | **62** | 348 | 27 | **62** | −660 | 26 |
| **64** | −629 | 28 | **64** | 50024 | 27 | **64** | 375 | 27 | **64** | −686 | 25 |
| **66** | 657 | 29 | **66** | 051 | 28 | **66** | −402 | 26 | **66** | 711 | 26 |
| **68** | −686 | 28 | **68** | −079 | 27 | **68** | 428 | 27 | **68** | 737 | 26 |
| **3070** | −714 | 28 | **3170** | −106 | 27 | **3270** | −455 | 26 | **3370** | −763 | 26 |
| **72** | 742 | 28 | **72** | 133 | 28 | **72** | 481 | 27 | **72** | −789 | 26 |
| **74** | 770 | 29 | **74** | −161 | 27 | **74** | −508 | 26 | **74** | −815 | 25 |
| **76** | −799 | 28 | **76** | 188 | 27 | **76** | 534 | 27 | **76** | 840 | 26 |
| **78** | −827 | 28 | **78** | 215 | 28 | **78** | −561 | 26 | **78** | −866 | 26 |
| **3080** | 855 | 28 | **3180** | −243 | 27 | **3280** | 587 | 27 | **3380** | −892 | 25 |
| **82** | 833 | 28 | **82** | 270 | 27 | **82** | −614 | 26 | **82** | 917 | 26 |
| **84** | 911 | 29 | **84** | 297 | 28 | **84** | 640 | 27 | **84** | 943 | 26 |
| **86** | −940 | 28 | **86** | −325 | 27 | **86** | −667 | 26 | **86** | −969 | 25 |
| **88** | −968 | 28 | **88** | −352 | 27 | **88** | 693 | 27 | **88** | 994 | 26 |
| **3090** | −996 | 28 | **3190** | 379 | 27 | **3290** | −720 | 26 | **3390** | 53020 | 26 |
| **92** | 49024 | 28 | **92** | 406 | 27 | **92** | −746 | 26 | **92** | −046 | 25 |
| **94** | 052 | 28 | **94** | 433 | 28 | **94** | 772 | 27 | **94** | 071 | 26 |
| **96** | 080 | 28 | **96** | −461 | 27 | **96** | −799 | 26 | **96** | −097 | 25 |
| **98** | 108 | 28 | **98** | −488 | 27 | **98** | 825 | 26 | **98** | 122 | 26 |
| **3100** | 136 | 28 | **3200** | −515 | 27 | **3300** | 851 | 27 | **3400** | −148 | 25 |
| **02** | 164 | 28 | **02** | 542 | 27 | **02** | −878 | 26 | **02** | 173 | 26 |
| **04** | 192 | 28 | **04** | 569 | 27 | **04** | 904 | 26 | **04** | −199 | 25 |
| **06** | 220 | 28 | **06** | 596 | 27 | **06** | 930 | 27 | **06** | 224 | 26 |
| **08** | 248 | 28 | **08** | 623 | 28 | **08** | −957 | 26 | **08** | −250 | 25 |
| **3110** | 276 | 28 | **3210** | −651 | 27 | **3310** | −983 | 26 | **3410** | 275 | 26 |
| **12** | −304 | 28 | **12** | −678 | 27 | **12** | 52009 | 26 | **12** | −301 | 25 |
| **14** | −332 | 28 | **14** | −705 | 27 | **14** | 035 | 26 | **14** | 326 | 26 |
| **16** | −360 | 28 | **16** | −732 | 27 | **16** | 061 | 27 | **16** | −352 | 25 |
| **18** | −388 | 27 | **18** | −759 | 27 | **18** | −088 | 26 | **18** | 377 | 26 |
| **3120** | 415 | 28 | **3220** | −786 | 27 | **3320** | −114 | 26 | **3420** | −403 | 25 |
| **22** | 443 | 28 | **22** | −813 | 27 | **22** | −140 | 26 | **22** | 428 | 25 |
| **24** | 471 | 28 | **24** | −840 | 26 | **24** | 166 | 26 | **24** | 453 | 26 |
| **26** | −499 | 28 | **26** | 866 | 27 | **26** | 192 | 26 | **26** | −479 | 25 |
| **28** | −527 | 27 | **28** | 893 | 27 | **28** | 218 | 26 | **28** | 504 | 25 |
| **3130** | 554 | 28 | **3230** | 920 | 27 | **3330** | 244 | 26 | **3430** | 529 | 26 |
| **32** | 582 | 28 | **32** | 947 | 27 | **32** | 270 | 27 | **32** | −555 | 25 |
| **34** | −610 | 28 | **34** | 974 | 27 | **34** | −297 | 26 | **34** | 580 | 25 |
| **36** | −638 | 27 | **36** | 51001 | 27 | **36** | −323 | 26 | **36** | 605 | 26 |
| **38** | 665 | 28 | **38** | −028 | 27 | **38** | −349 | 26 | **38** | −631 | 25 |
| **3140** | −693 | 28 | **3240** | −055 | 26 | **3340** | −375 | 26 | **3440** | −656 | 25 |
| **42** | −721 | 27 | **42** | 081 | 27 | **42** | −401 | 26 | **42** | 681 | 25 |
| **44** | 748 | 28 | **44** | 108 | 27 | **44** | −427 | 26 | **44** | 706 | 26 |
| **46** | −776 | 27 | **46** | −135 | .27 | **46** | −453 | 26 | **46** | −732 | 25 |
| **48** | 803 | 28 | **48** | −162 | 26 | **48** | −479 | 25 | **48** | −757 | 25 |

P. P.

| | **29** | **28** | **27** | **26** | **25** |
|---|---|---|---|---|---|
| **1** | 1 | 1 | 1 | 1 | 1 |
| **2** | 3 | 3 | 3 | 3 | 3 |
| **3** | 4 | 4 | 4 | 4 | 4 |
| **4** | 6 | 6 | 5 | 5 | 5 |
| **5** | 7 | 7 | 7 | 7 | 6 |
| **6** | 9 | 8 | 8 | 8 | 8 |
| **7** | 10 | 10 | 9 | 9 | 9 |
| **8** | 12 | 11 | 11 | 10 | 10 |
| **9** | 13 | 13 | 12 | 12 | 11 |
| **10** | 15 | 14 | 14 | 13 | 13 |

## LOGARITHMS OF NUMBERS.—*Continued.*

| No. | Log. | Dif. | No. | Log. | Dif. | No. | Log. | Dif. | No. | Log. | Dif. |
|---|---|---|---|---|---|---|---|---|---|---|---|
| **3450** | 53782 | 25 | **3550** | 55023 | 24 | **3650** | 56229 | 24 | **3750** | 57403 | 58 |
| **52** | 807 | 25 | **52** | 047 | 25 | **52** | 253 | 24 | **55** | −461 | 58 |
| **54** | 832 | 25 | **54** | —072 | 24 | **54** | —277 | 24 | **60** | −519 | 57 |
| **56** | 857 | 25 | **56** | 096 | 25 | **56** | —301 | 23 | **65** | 576 | 58 |
| **58** | 882 | 26 | **58** | —121 | 24 | **58** | 324 | 24 | **70** | 634 | 58 |
| **3460** | —908 | 24 | **3560** | —145 | 24 | **3660** | 348 | 24 | **3775** | −692 | 57 |
| **62** | —933 | 25 | **62** | 169 | 25 | **62** | —372 | 24 | **80** | 749 | 58 |
| **64** | —958 | 24 | **64** | —194 | 24 | **64** | —396 | 23 | **85** | −807 | 57 |
| **66** | —983 | 24 | **66** | 218 | 24 | **66** | 419 | 24 | **90** | −864 | 57 |
| **68** | 54008 | 25 | **68** | 242 | 25 | **68** | —443 | 24 | **95** | 921 | 57 |
| **3470** | —033 | 25 | **3570** | —267 | 24 | **3670** | —467 | 23 | **3800** | 978 | 57 |
| **72** | —058 | 25 | **72** | 291 | 24 | **72** | 490 | 24 | **05** | 58035 | 57 |
| **74** | —083 | 25 | **74** | 315 | 25 | **74** | —514 | 24 | **10** | 092 | 57 |
| **76** | —108 | 25 | **76** | —340 | 24 | **76** | —538 | 23 | **15** | 149 | 57 |
| **78** | —133 | 25 | **78** | 364 | 24 | **78** | 561 | 24 | **20** | 206 | 57 |
| **3480** | —158 | 25 | **3580** | 388 | 25 | **3680** | —585 | 23 | **3825** | 263 | 57 |
| **82** | —183 | 25 | **82** | —413 | 24 | **82** | 608 | 24 | **30** | −320 | 57 |
| **84** | —208 | 25 | **84** | —437 | 24 | **84** | —632 | 24 | **35** | −377 | 56 |
| **86** | —233 | 25 | **86** | 461 | 24 | **86** | —656 | 23 | **40** | 433 | 57 |
| **88** | —258 | 25 | **88** | 485 | 24 | **88** | 679 | 24 | **45** | −490 | 56 |
| **3490** | —283 | 24 | **3590** | 509 | 25 | **3690** | —703 | 23 | **3850** | 546 | 56 |
| **92** | 307 | 25 | **92** | —534 | 24 | **92** | 726 | 24 | **55** | 602 | 57 |
| **94** | 332 | 25 | **94** | —558 | 24 | **94** | —750 | 23 | **60** | −659 | 56 |
| **96** | 357 | 25 | **96** | —582 | 24 | **96** | 773 | 24 | **65** | −715 | 56 |
| **98** | —382 | 25 | **98** | 606 | 24 | **98** | —797 | 23 | **70** | 771 | 56 |
| **3500** | —407 | 25 | **3600** | 630 | 24 | **3700** | 820 | 24 | **3875** | 827 | 56 |
| **02** | —432 | 24 | **02** | 654 | 24 | **02** | —844 | 23 | **80** | 883 | 56 |
| **04** | 456 | 25 | **04** | 678 | 25 | **04** | 867 | 24 | **85** | 939 | 56 |
| **06** | 481 | 25 | **06** | —703 | 24 | **06** | —891 | 23 | **90** | −995 | 56 |
| **08** | —506 | 25 | **08** | —727 | 24 | **08** | −914 | 23 | **95** | 59051 | 55 |
| **3510** | —531 | 24 | **3610** | —751 | 24 | **3710** | 937 | 24 | **3900** | 106 | 56 |
| **12** | 555 | 25 | **12** | —775 | 24 | **12** | −961 | 23 | **05** | 162 | 56 |
| **14** | 580 | 25 | **14** | —799 | 24 | **14** | 984 | 24 | **10** | −218 | 55 |
| **16** | −605 | 25 | **16** | —823 | 24 | **16** | 57008 | 23 | **15** | 273 | 56 |
| **18** | —630 | 24 | **18** | —847 | 24 | **18** | −031 | 23 | **20** | −329 | 55 |
| **3520** | 654 | 25 | **3620** | —871 | 24 | **3720** | 054 | 24 | **3925** | −384 | 55 |
| **22** | —679 | 25 | **22** | —895 | 24 | **22** | −078 | 23 | **30** | 439 | 55 |
| **24** | —704 | 24 | **24** | —919 | 24 | **24** | −101 | 23 | **35** | 494 | 56 |
| **26** | 728 | 25 | **26** | —943 | 24 | **26** | 124 | 24 | **40** | −550 | 55 |
| **28** | —753 | 24 | **28** | —967 | 24 | **28** | −148 | 23 | **45** | −605 | 55 |
| **3530** | 777 | 25 | **3630** | —991 | 24 | **3730** | −171 | 23 | **3950** | −660 | 55 |
| **32** | 802 | 25 | **32** | 56015 | 23 | **32** | 194 | 23 | **55** | −715 | 55 |
| **34** | —827 | 24 | **34** | 038 | 24 | **34** | 217 | 24 | **60** | −770 | 54 |
| **36** | 851 | 25 | **36** | 062 | 24 | **36** | −241 | 23 | **65** | 824 | 55 |
| **38** | —876 | 24 | **38** | 086 | 24 | **38** | −264 | 23 | **70** | 879 | 55 |
| **3540** | 900 | 25 | **3640** | 110 | 24 | **3740** | 287 | 23 | **3975** | −934 | 54 |
| **42** | —925 | 24 | **42** | —134 | 24 | **42** | 310 | 24 | **80** | 988 | 55 |
| **44** | 949 | 25 | **44** | —158 | 24 | **44** | −334 | 23 | **85** | 60043 | 54 |
| **46** | —974 | 24 | **46** | —182 | 23 | **46** | −357 | 23 | **90** | 097 | 55 |
| **48** | 998 | 25 | **48** | 205 | 24 | **48** | −380 | 23 | **95** | −152 | 54 |

P. P.

| | **26** | **25** |
|---|---|---|
| **1** | 1 | 1 |
| **2** | 3 | 3 |
| **3** | 4 | 4 |
| **4** | 5 | 5 |
| **5** | 7 | 6 |
| **6** | 8 | 8 |
| **7** | 9 | 9 |
| **8** | 10 | 10 |
| **9** | 12 | 11 |
| **10** | 13 | 13 |

| | **24** | **23** |
|---|---|---|
| **1** | 1 | 1 |
| **2** | 2 | 2 |
| **3** | 4 | 3 |
| **4** | 5 | 5 |
| **5** | 6 | 6 |
| **6** | 7 | 7 |
| **7** | 8 | 8 |
| **8** | 10 | 9 |
| **9** | 11 | 10 |
| **10** | 12 | 12 |

| | **58** | **57** |
|---|---|---|
| **1** | 1.2 | 1.1 |
| **2** | 2.3 | 2.3 |
| **3** | 3.5 | 3.4 |
| **4** | 4.6 | 4.6 |
| **5** | 5.8 | 5.7 |
| **6** | 7.0 | 6.8 |
| **7** | 8.1 | 8.0 |
| **8** | 9.3 | 9.1 |
| **9** | 10.4 | 10.3 |
| **10** | 11.6 | 11.4 |

| | **56** | **55** |
|---|---|---|
| **1** | 1.1 | 1.1 |
| **2** | 2.2 | 2.2 |
| **3** | 3.4 | 3.3 |
| **4** | 4.5 | 4.4 |
| **5** | 5.6 | 5.5 |
| **6** | 6.7 | 6.6 |
| **7** | 7.8 | 7.7 |
| **8** | 9.0 | 8.8 |
| **9** | 10.1 | 9.9 |
| **10** | 11.2 | 11.0 |

| | **54** |
|---|---|
| **1** | 1.1 |
| **2** | 2.2 |
| **3** | 3.2 |
| **4** | 4.3 |
| **5** | 5.4 |
| **6** | 6.5 |
| **7** | 7.6 |
| **8** | 8.6 |
| **9** | 9.7 |
| **10** | 10.8 |

## LOGARITHMS OF NUMBERS.—*Continued.*

| No. | Log | Dif. | No. | Log. | Dif. | No. | Log. | Dif. | No. | Log. | Dif. | P. P. | | | |
|---|---|---|---|---|---|---|---|---|---|---|---|---|---|---|---|
| **4000** | 60206 | 54 | **4250** | 62839 | 51 | **4500** | 65321 | 48 | **4750** | 67669 | 46 | | **54** | **53** | **52** |
| **05** | 260 | 54 | **55** | -890 | 51 | **05** | 369 | 49 | **55** | 715 | 46 | **1** | 1.1 | 1.1 | 1.0 |
| **10** | 314 | 55 | **60** | -941 | 51 | **10** | -418 | 48 | **60** | -761 | 45 | **2** | 2.2 | 2.1 | 2.1 |
| **15** | -369 | 54 | **65** | -992 | 51 | **15** | -466 | 48 | **65** | 806 | 46 | **3** | 3.2 | 3.2 | 3.1 |
| **20** | -423 | 54 | **70** | 63043 | 51 | **20** | -514 | 48 | **70** | -852 | 45 | **4** | 4.3 | 4.2 | 4.2 |
| | | | | | | | | | | | | **5** | 5.4 | 5.3 | 5.2 |
| **25** | -477 | 54 | **75** | -094 | 50 | **25** | -562 | 48 | **75** | 897 | 46 | **6** | 6.5 | 6.4 | 6.2 |
| **30** | -531 | 53 | **80** | 144 | 51 | **30** | -610 | 48 | **80** | -943 | 45 | **7** | 7.6 | 7.4 | 7.3 |
| **35** | 584 | 54 | **85** | 195 | 51 | **35** | -658 | 48 | **85** | 968 | 46 | **8** | 8.6 | 8.5 | 8.3 |
| **40** | 638 | 54 | **90** | -246 | 50 | **40** | -706 | 47 | **90** | 68034 | 45 | **9** | 9.7 | 9.5 | 9.4 |
| **45** | -692 | 54 | **95** | 296 | 51 | **45** | 753 | 48 | **95** | -079 | 45 | **10** | 10.8 | 10.6 | 10.4 |
| **4050** | -746 | 53 | **4300** | -347 | 50 | **4550** | 801 | 48 | **4800** | 124 | 45 | | **51** | **50** | **49** |
| **55** | 799 | 54 | **05** | 397 | 51 | **55** | -849 | 47 | **05** | 169 | 46 | **1** | 1.0 | 1.0 | 1.0 |
| **60** | -853 | 53 | **10** | -448 | 50 | **60** | 896 | 48 | **10** | -215 | 45 | **2** | 2.0 | 2.0 | 2.0 |
| **65** | 906 | 53 | **15** | 498 | 50 | **65** | 944 | 48 | **15** | -260 | 45 | **3** | 3.1 | 3.0 | 2.9 |
| **70** | 959 | 54 | **20** | 548 | 51 | **70** | -992 | 47 | **20** | -305 | 45 | **4** | 4.1 | 4.0 | 3.9 |
| | | | | | | | | | | | | **5** | 5.1 | 5.0 | 4.9 |
| **75** | 61013 | 53 | **25** | -599 | 50 | **75** | 66039 | 48 | **25** | -350 | 45 | **6** | 6.1 | 6.0 | 5.9 |
| **80** | 066 | 53 | **30** | -649 | 50 | **80** | -087 | 47 | **30** | -395 | 45 | **7** | 7.1 | 7.0 | 6.9 |
| **85** | 199 | 53 | **35** | -699 | 50 | **85** | -134 | 47 | **35** | -440 | 45 | **8** | 8.2 | 8.0 | 7.8 |
| **90** | 172 | 53 | **40** | -749 | 50 | **90** | 181 | 48 | **40** | -485 | 44 | **9** | 9.2 | 9.0 | 8.8 |
| **95** | 225 | 53 | **45** | -799 | 50 | **95** | -229 | 47 | **45** | 529 | 45 | **10** | 10.2 | 10.0 | 9.8 |
| **4100** | 278 | 53 | **4350** | -849 | 50 | **4600** | -276 | 47 | **4850** | 574 | 45 | | **48** | **47** | |
| **05** | 331 | 53 | **55** | -899 | 50 | **05** | -323 | 47 | **55** | -619 | 45 | **1** | 1.0 | 0.9 | |
| **10** | 384 | 53 | **60** | -949 | 49 | **10** | 370 | 47 | **60** | -664 | 44 | **2** | 1.9 | 1.9 | |
| **15** | -437 | 53 | **65** | 998 | 50 | **15** | 417 | 47 | **65** | 708 | 45 | **3** | 2.9 | 2.8 | |
| **20** | -490 | 52 | **70** | 64048 | 50 | **20** | 464 | 47 | **70** | -753 | 44 | **4** | 3.8 | 3.8 | |
| | | | | | | | | | | | | **5** | 4.8 | 4.7 | |
| **25** | 542 | 53 | **75** | -098 | 49 | **25** | 511 | 47 | **75** | 797 | 45 | **6** | 5.8 | 5.6 | |
| **30** | 595 | 53 | **80** | 147 | 50 | **30** | 558 | 47 | **80** | -842 | 44 | **7** | 6.7 | 6.6 | |
| **35** | -648 | 52 | **85** | -197 | 49 | **35** | -605 | 47 | **85** | 886 | 45 | **8** | 7.7 | 7.5 | |
| **40** | 700 | 52 | **90** | 245 | 50 | **40** | -652 | 47 | **90** | -931 | 44 | **9** | 8.6 | 8.5 | |
| **45** | 752 | 53 | **95** | -296 | 49 | **45** | -699 | 46 | **95** | 975 | 45 | **10** | 9.6 | 9.4 | |
| **4150** | -805 | 52 | **4400** | 345 | 50 | **4650** | 745 | 47 | **4900** | 69020 | 44 | | **46** | **45** | |
| **55** | 857 | 52 | **05** | -395 | 49 | **55** | -792 | 47 | **05** | -064 | 44 | **1** | 0.9 | 0.9 | |
| **60** | 909 | 53 | **10** | -444 | 49 | **60** | -839 | 46 | **10** | 108 | 44 | **2** | 1.8 | 1.8 | |
| **65** | -962 | 52 | **15** | 493 | 49 | **65** | 885 | 47 | **15** | 152 | 45 | **3** | 2.8 | 2.7 | |
| **70** | 62014 | 52 | **20** | 542 | 49 | **70** | -932 | 46 | **20** | -197 | 44 | **4** | 3.7 | 3.6 | |
| | | | | | | | | | | | | **5** | 4.6 | 4.5 | |
| **75** | -066 | 52 | **25** | 591 | 49 | **75** | 978 | 47 | **25** | -241 | 44 | **6** | 5.5 | 5.4 | |
| **80** | -118 | 52 | **30** | 640 | 49 | **80** | 67025 | 46 | **30** | -285 | 44 | **7** | 6.4 | 6.3 | |
| **85** | -170 | 51 | **35** | 689 | 49 | **85** | -071 | 46 | **35** | -329 | 44 | **8** | 7.4 | 7.2 | |
| **90** | 221 | 52 | **40** | 738 | 49 | **90** | 117 | 47 | **40** | -373 | 44 | **9** | 8.3 | 8.1 | |
| **95** | 273 | 52 | **45** | 787 | 49 | **95** | -164 | 46 | **45** | -417 | 44 | **10** | 9.2 | 9.0 | |
| **4200** | -325 | 52 | **4450** | 836 | 49 | **4700** | -210 | 46 | **4950** | -461 | 43 | | **44** | **43** | |
| **05** | -377 | 51 | **55** | -855 | 48 | **05** | -256 | 46 | **55** | 504 | 44 | **1** | 0.9 | 0.9 | |
| **10** | 428 | 52 | **60** | 933 | 49 | **10** | 302 | 46 | **60** | 548 | 44 | **2** | 1.8 | 1.7 | |
| **15** | -480 | 51 | **65** | 982 | 49 | **15** | 348 | 46 | **65** | -592 | 44 | **3** | 2.6 | 2.6 | |
| **20** | 531 | 52 | **70** | 65031 | 48 | **20** | 394 | 45 | **70** | -636 | 43 | **4** | 3.5 | 3.4 | |
| | | | | | | | | | | | | **5** | 4.4 | 4.3 | |
| **25** | -583 | 51 | **75** | 079 | 49 | **25** | 440 | 46 | **75** | 679 | 44 | **6** | 5.3 | 5.2 | |
| **30** | 634 | 51 | **80** | -128 | 48 | **30** | 486 | 46 | **80** | -723 | 44 | **7** | 6.2 | 6.0 | |
| **35** | 685 | 52 | **85** | 176 | 49 | **35** | -532 | 46 | **85** | -767 | 43 | **8** | 7.0 | 6.9 | |
| **40** | -737 | 51 | **90** | -225 | 48 | **40** | -578 | 46 | **90** | 810 | 44 | **9** | 7.9 | 7.7 | |
| **45** | -788 | 51 | **95** | -273 | 48 | **45** | -624 | 45 | **95** | -854 | 43 | **10** | 8.8 | 8.6 | |

## LOGARITHMS OF NUMBERS.—*Continued.*

| No. | Log. | Dif. | No. | Log. | Dif. | No. | Log. | Dif. | No. | Log. | Dif. |
|---|---|---|---|---|---|---|---|---|---|---|---|
| **5000** | 69897 | 43 | **5250** | 72016 | 41 | **5500** | 74036 | 40 | **5750** | 75967 | 38 |
| **05** | 940 | 44 | **55** | 057 | 42 | **05** | −076 | 39 | **55** | 76005 | 37 |
| **10** | −984 | 43 | **60** | −099 | 41 | **10** | 115 | 40 | **60** | 042 | 38 |
| **15** | 70027 | 43 | **65** | −140 | 41 | **15** | −155 | 39 | **65** | −080 | 38 |
| **20** | 070 | 44 | **70** | 181 | 41 | **20** | −194 | 39 | **70** | −118 | 37 |
| **25** | −114 | 43 | **75** | 222 | 41 | **25** | 233 | 40 | **75** | 155 | 38 |
| **30** | −157 | 43 | **80** | 263 | 41 | **30** | −273 | 39 | **80** | −193 | 37 |
| **35** | −200 | 43 | **85** | 304 | 42 | **35** | −312 | 39 | **85** | 230 | 38 |
| **40** | 243 | 43 | **90** | −346 | 41 | **40** | −351 | 39 | **90** | −268 | 37 |
| **45** | 286 | 43 | **95** | −387 | 41 | **45** | 390 | 39 | **95** | 305 | 38 |
| **5050** | 329 | 43 | **5300** | −428 | 41 | **5550** | 429 | 39 | **5800** | −343 | 37 |
| **55** | 372 | 43 | **05** | −469 | 40 | **55** | 468 | 39 | **05** | 380 | 38 |
| **60** | 415 | 43 | **10** | 509 | 41 | **60** | 507 | 40 | **10** | −418 | 37 |
| **65** | −458 | 43 | **15** | 550 | 41 | **65** | −547 | 39 | **15** | −455 | 37 |
| **70** | −501 | 43 | **20** | 591 | 41 | **70** | −586 | 38 | **20** | 492 | 38 |
| **75** | −544 | 42 | **25** | −632 | 41 | **75** | 624 | 39 | **25** | −530 | 37 |
| **80** | 586 | 43 | **30** | −673 | 40 | **80** | 663 | 39 | **30** | −567 | 37 |
| **85** | 629 | 43 | **35** | 713 | 41 | **85** | 702 | 39 | **35** | 604 | 37 |
| **90** | −672 | 42 | **40** | 754 | 41 | **90** | 741 | 39 | **40** | 641 | 37 |
| **95** | 714 | 43 | **45** | −795 | 40 | **95** | 780 | 39 | **45** | 678 | 38 |
| **5100** | 757 | 43 | **5350** | 835 | 41 | **5600** | −819 | 39 | **5850** | −716 | 37 |
| **05** | −800 | 42 | **55** | −876 | 40 | **05** | −858 | 38 | **55** | −753 | 37 |
| **10** | 842 | 43 | **60** | 916 | 41 | **10** | 896 | 39 | **60** | −790 | 37 |
| **15** | −885 | 42 | **65** | −957 | 40 | **15** | −935 | 39 | **65** | −827 | 37 |
| **20** | −927 | 42 | **70** | 997 | 41 | **20** | −974 | 38 | **70** | −864 | 37 |
| **25** | 969 | 43 | **75** | 73038 | 40 | **25** | 75012 | 39 | **75** | −901 | 37 |
| **30** | 71012 | 42 | **80** | 078 | 41 | **30** | −051 | 38 | **80** | −938 | 37 |
| **35** | 054 | 42 | **85** | −119 | 40 | **35** | 089 | 39 | **85** | −975 | 37 |
| **40** | 096 | 43 | **90** | −159 | 40 | **40** | −128 | 38 | **90** | 77012 | 36 |
| **45** | −139 | 42 | **95** | 199 | 40 | **45** | 166 | 39 | **95** | 048 | 37 |
| **5150** | −181 | 42 | **5400** | 239 | 41 | **5650** | −205 | 38 | **5900** | 085 | 37 |
| **55** | −223 | 42 | **05** | −280 | 40 | **55** | 243 | 39 | **05** | −122 | 37 |
| **60** | −265 | 42 | **10** | −320 | 40 | **60** | −282 | 38 | **10** | −159 | 36 |
| **65** | 307 | 42 | **15** | −360 | 40 | **65** | −320 | 38 | **15** | 195 | 37 |
| **70** | 349 | 42 | **20** | −400 | 40 | **70** | 358 | 39 | **20** | 232 | 37 |
| **75** | 391 | 42 | **25** | −440 | 40 | **75** | −397 | 38 | **25** | −269 | 36 |
| **80** | −433 | 42 | **30** | −480 | 40 | **80** | −435 | 38 | **30** | 305 | 37 |
| **85** | −475 | 42 | **35** | −520 | 40 | **85** | 473 | 38 | **35** | 342 | 37 |
| **90** | −517 | 42 | **40** | −560 | 40 | **90** | 511 | 38 | **40** | −379 | 36 |
| **95** | −559 | 41 | **45** | −600 | 40 | **95** | 549 | 38 | **45** | 415 | 37 |
| **5200** | 600 | 42 | **5450** | −640 | 39 | **5700** | 587 | 39 | **5950** | −452 | 36 |
| **05** | 642 | 42 | **55** | 679 | 40 | **05** | −626 | 38 | **55** | 488 | 37 |
| **10** | −684 | 41 | **60** | 719 | 40 | **10** | −664 | 38 | **60** | −525 | 36 |
| **15** | 725 | 42 | **65** | 759 | 40 | **15** | −702 | 38 | **65** | 561 | 36 |
| **20** | 767 | 42 | **70** | −799 | 39 | **20** | −740 | 38 | **70** | 597 | 37 |
| **25** | −809 | 41 | **75** | 838 | 40 | **25** | −778 | 37 | **75** | −634 | 36 |
| **30** | 850 | 42 | **80** | 878 | 40 | **30** | 815 | 38 | **80** | 670 | 36 |
| **35** | −892 | 41 | **85** | −918 | 39 | **35** | 853 | 38 | **C5** | 706 | 37 |
| **40** | 933 | 42 | **90** | 957 | 40 | **40** | 891 | 38 | **90** | −743 | 36 |
| **45** | −975 | 41 | **95** | −997 | 39 | **45** | 929 | 38 | **95** | −779 | 36 |

**P. P.**

| | **44** | **43** |
|---|---|---|
| **1** | 0.9 | 0.9 |
| **2** | 1.8 | 1.7 |
| **3** | 2.6 | 2.6 |
| **4** | 3.5 | 3.4 |
| **5** | 4.4 | 4.3 |
| **6** | 5.3 | 5.2 |
| **7** | 6.2 | 6.0 |
| **8** | 7.0 | 6.9 |
| **9** | 7.9 | 7.7 |
| **10** | 8.8 | 8.6 |

| | **42** | **41** |
|---|---|---|
| **1** | 0.8 | 0.8 |
| **2** | 1.7 | 1.6 |
| **3** | 2.5 | 2.5 |
| **4** | 3.4 | 3.3 |
| **5** | 4.2 | 4.1 |
| **6** | 5.0 | 4.9 |
| **7** | 5 9 | 5.7 |
| **8** | 6.7 | 6.6 |
| **9** | 7.6 | 7.4 |
| **10** | 8.4 | 8.2 |

| | **40** | **39** |
|---|---|---|
| **1** | 0.8 | 0.8 |
| **2** | 1.6 | 1.6 |
| **3** | 2.4 | 2.3 |
| **4** | 3.2 | 3.1 |
| **5** | 4.0 | 3.9 |
| **6** | 4.8 | 4.7 |
| **7** | 5.6 | 5.5 |
| **8** | 6.4 | 6.2 |
| **9** | 7.2 | 7.0 |
| **10** | 8.0 | 7.8 |

| | **38** | **37** |
|---|---|---|
| **1** | 0.8 | 0.7 |
| **2** | 1.5 | 1.5 |
| **3** | 2.3 | 2.2 |
| **4** | 3.0 | 3.0 |
| **5** | 3.8 | 3.7 |
| **6** | 4.6 | 4.4 |
| **7** | 5.3 | 5.2 |
| **8** | 6.1 | 5.9 |
| **9** | 6.8 | 6.7 |
| **10** | 7.6 | 7.4 |

| | **36** |
|---|---|
| **1** | 0.7 |
| **2** | 1.4 |
| **3** | 2.2 |
| **4** | 2.9 |
| **5** | 3.6 |
| **6** | 4.3 |
| **7** | 5.0 |
| **8** | 5.8 |
| **9** | 6.5 |
| **10** | 7.2 |

## LOGARITHMS OF NUMBERS.—*Continued.*

| No. | Log. | Dif. | No. | Log. | Dif. | No. | Log. | Dif. | No. | Log. | Dif. |
|---|---|---|---|---|---|---|---|---|---|---|---|
| 6000 | 77815 | 36 | 6250 | 79588 | 35 | 6500 | 81291 | 34 | 6750 | 82930 | 33 |
| 05 | 851 | 36 | 55 | −623 | 34 | 05 | −325 | 33 | 55 | −963 | 32 |
| 10 | 887 | 37 | 60 | 657 | 35 | 10 | 358 | 33 | 60 | −995 | 32 |
| 15 | −924 | 36 | 65 | 692 | 35 | 15 | 391 | 34 | 65 | 83027 | 32 |
| 20 | −960 | 36 | 70 | −727 | 34 | 20 | −425 | 33 | 70 | −059 | 32 |
| 25 | −996 | 36 | 75 | 761 | 35 | 25 | 458 | 33 | 75 | −091 | 32 |
| 30 | 78032 | 36 | 80 | −796 | 35 | 30 | 491 | 34 | 80 | −123 | 32 |
| 35 | −068 | 36 | 85 | −831 | 34 | 35 | −525 | 33 | 85 | −155 | 32 |
| 40 | −104 | 36 | 90 | 865 | 35 | 40 | −558 | 33 | 90 | −187 | 32 |
| 45 | −140 | 36 | 95 | −900 | 34 | 45 | −591 | 33 | 95 | −219 | 32 |
| 6050 | −176 | 35 | 6300 | 934 | 35 | 6550 | 624 | 33 | 6800 | −251 | 32 |
| 55 | 211 | 36 | 05 | −969 | 34 | 55 | 657 | 33 | 05 | −283 | 32 |
| 60 | 247 | 36 | 10 | 80003 | 34 | 60 | 690 | 33 | 10 | −315 | 32 |
| 65 | 283 | 36 | 15 | 037 | 35 | 65 | 723 | 34 | 15 | −347 | 31 |
| 70 | −319 | 36 | 20 | −072 | 34 | 70 | −757 | 33 | 20 | 378 | 32 |
| 75 | −355 | 35 | 25 | 106 | 34 | 75 | −790 | 33 | 25 | 410 | 32 |
| 80 | 390 | 36 | 30 | 140 | 35 | 80 | −823 | 33 | 30 | 442 | 32 |
| 85 | 426 | 36 | 35 | −175 | 34 | 85 | −856 | 33 | 35 | −474 | 32 |
| 90 | −462 | 35 | 40 | −209 | 34 | 90 | −889 | 32 | 40 | −506 | 31 |
| 95 | 497 | 36 | 45 | 243 | 34 | 95 | 921 | 33 | 45 | 537 | 32 |
| 6100 | −533 | 36 | 6350 | 277 | 35 | 6600 | 954 | 33 | 6850 | 569 | 32 |
| 05 | −569 | 35 | 55 | −312 | 34 | 05 | 987 | 33 | 55 | −601 | 31 |
| 10 | 604 | 36 | 60 | −346 | 34 | 10 | 82020 | 33 | 60 | 632 | 32 |
| 15 | −640 | 35 | 65 | −380 | 34 | 15 | −053 | 33 | 65 | 664 | 32 |
| 20 | 675 | 36 | 70 | −414 | 34 | 20 | −086 | 33 | 70 | −696 | 31 |
| 25 | −711 | 35 | 75 | 448 | 34 | 25 | −119 | 32 | 75 | 727 | 32 |
| 30 | 746 | 35 | 80 | 482 | 34 | 30 | 151 | 33 | 80 | −759 | 31 |
| 35 | 781 | 36 | 85 | 516 | 34 | 35 | 184 | 33 | 85 | 790 | 32 |
| 40 | −817 | 35 | 90 | 550 | 34 | 40 | −217 | 32 | 90 | −822 | 31 |
| 45 | 852 | 36 | 95 | 584 | 34 | 45 | 249 | 33 | 95 | 853 | 32 |
| 6150 | −888 | 35 | 6400 | −618 | 34 | 6650 | 282 | 33 | 6900 | −885 | 31 |
| 55 | −923 | 35 | 05 | −652 | 34 | 55 | −315 | 32 | 05 | 916 | 32 |
| 60 | 958 | 35 | 10 | −686 | 34 | 60 | 347 | 33 | 10 | −948 | 31 |
| 65 | 993 | 36 | 15 | −720 | 34 | 65 | 380 | 33 | 15 | 979 | 32 |
| 70 | 79029 | 35 | 20 | −754 | 33 | 70 | −413 | 32 | 20 | 84011 | 31 |
| 75 | −064 | 35 | 25 | 787 | 34 | 75 | 445 | 33 | 25 | −042 | 31 |
| 80 | −099 | 35 | 30 | 821 | 34 | 80 | −478 | 32 | 30 | 073 | 32 |
| 85 | −134 | 35 | 35 | −855 | 34 | 85 | 510 | 33 | 35 | −105 | 31 |
| 90 | 169 | 35 | 40 | −889 | 33 | 90 | −543 | 32 | 40 | −136 | 31 |
| 95 | 204 | 35 | 45 | 922 | 34 | 95 | 575 | 32 | 45 | 167 | 31 |
| 6200 | 239 | 35 | 6450 | −956 | 34 | 6700 | 607 | 33 | 6950 | 198 | 32 |
| 05 | 274 | 35 | 55 | −990 | 33 | 05 | −640 | 32 | 55 | −230 | 31 |
| 10 | 309 | 35 | 60 | 81023 | 34 | 10 | 672 | 33 | 60 | −261 | 31 |
| 15 | 344 | 35 | 65 | −057 | 33 | 15 | −705 | 32 | 65 | 292 | 31 |
| 20 | 379 | 35 | 70 | 090 | 34 | 20 | −737 | 32 | 70 | 323 | 31 |
| 25 | −414 | 35 | 75 | −124 | 34 | 25 | 769 | 33 | 75 | 354 | 32 |
| 30 | −449 | 35 | 80 | −158 | 33 | 30 | −802 | 32 | 80 | −386 | 31 |
| 35 | −484 | 34 | 85 | −191 | 33 | 35 | −834 | 32 | 85 | −417 | 31 |
| 40 | 518 | 35 | 90 | 224 | 34 | 40 | −866 | 32 | 90 | −448 | 31 |
| 45 | 553 | 35 | 95 | −258 | 33 | 45 | 898 | 32 | 95 | −479 | 31 |

**P. P.**

| | 37 | 36 |
|---|---|---|
| 1 | 0.7 | 0.7 |
| 2 | 1.5 | 1.4 |
| 3 | 2.2 | 2.2 |
| 4 | 3.0 | 2.9 |
| 5 | 3.7 | 3.6 |
| 6 | 4.4 | 4.3 |
| 7 | 5.2 | 5.0 |
| 8 | 5.9 | 5.8 |
| 9 | 6.7 | 6.5 |
| 10 | 7.4 | 7.2 |

| | 35 | 34 |
|---|---|---|
| 1 | 0.7 | 0.7 |
| 2 | 1.4 | 1.4 |
| 3 | 2.1 | 2.0 |
| 4 | 2.8 | 2.7 |
| 5 | 3.5 | 3.4 |
| 6 | 4.2 | 4.1 |
| 7 | 4.9 | 4.8 |
| 8 | 5.6 | 5.4 |
| 9 | 6.3 | 6.1 |
| 10 | 7.0 | 6.8 |

| | 33 |
|---|---|
| 1 | 0.7 |
| 2 | 1.3 |
| 3 | 2.0 |
| 4 | 2.6 |
| 5 | 3.3 |
| 6 | 4.0 |
| 7 | 4.6 |
| 8 | 5.3 |
| 9 | 5.9 |
| 10 | 6.6 |

| | 32 |
|---|---|
| 1 | 0.6 |
| 2 | 1.3 |
| 3 | 1.9 |
| 4 | 2.6 |
| 5 | 3.2 |
| 6 | 3.8 |
| 7 | 4.5 |
| 8 | 5.1 |
| 9 | 5.8 |
| 10 | 6.4 |

| | 31 |
|---|---|
| 1 | 0.6 |
| 2 | 1.2 |
| 3 | 1.9 |
| 4 | 2.5 |
| 5 | 3.1 |
| 6 | 3.7 |
| 7 | 4.3 |
| 8 | 5.0 |
| 9 | 5.6 |
| 10 | 6.2 |

## LOGARITHMS OF NUMBERS. — *Continued.*

| No. | Log. | Dif. | No. | Log. | Dif. | No. | Log. | Dif. | No. | Log. | Dif. |
|---|---|---|---|---|---|---|---|---|---|---|---|
| **7000** | 84510 | 31 | **7250** | 86034 | 30 | **7500** | 87506 | 29 | **7750** | 88930 | 28 |
| **05** | −541 | 31 | **55** | −064 | 30 | **05** | 535 | 29 | **55** | 958 | 28 |
| **10** | −572 | 31 | **60** | −094 | 30 | **10** | −564 | 29 | **60** | 986 | 28 |
| **15** | −603 | 31 | **65** | −124 | 29 | **15** | −593 | 29 | **65** | 89014 | 28 |
| **20** | −634 | 31 | **70** | 153 | 30 | **20** | −622 | 29 | **70** | 042 | 28 |
| **25** | −665 | 31 | **75** | 183 | 30 | **25** | −651 | 28 | **75** | 070 | 28 |
| **30** | −696 | 30 | **80** | 213 | 30 | **30** | 679 | 29 | **80** | −098 | 28 |
| **35** | 726 | 31 | **85** | −243 | 30 | **35** | 708 | 29 | **85** | −126 | 28 |
| **40** | 757 | 31 | **90** | −273 | 30 | **40** | 737 | 29 | **90** | −154 | 28 |
| **45** | 788 | 31 | **95** | −303 | 29 | **45** | −766 | 29 | **95** | −182 | 27 |
| **7050** | −819 | 31 | **7300** | 332 | 30 | **7550** | −795 | 28 | **7800** | 209 | 28 |
| **55** | −850 | 30 | **05** | 362 | 30 | **55** | 823 | 29 | **05** | 237 | 28 |
| **60** | 880 | 31 | **10** | −392 | 29 | **60** | 852 | 29 | **10** | 265 | 28 |
| **65** | 911 | 31 | **15** | 421 | 30 | **65** | −881 | 29 | **15** | −293 | 28 |
| **70** | −942 | 31 | **20** | 451 | 30 | **70** | −910 | 28 | **20** | −321 | 27 |
| **75** | −973 | 30 | **25** | −481 | 29 | **75** | 938 | 29 | **25** | 348 | 28 |
| **80** | 85003 | 31 | **30** | 510 | 30 | **80** | −967 | 29 | **30** | 376 | 28 |
| **85** | −034 | 31 | **35** | 540 | 30 | **85** | −996 | 28 | **35** | −404 | 28 |
| **90** | −065 | 30 | **40** | −570 | 29 | **90** | 88024 | 29 | **40** | −432 | 27 |
| **95** | 095 | 31 | **45** | 599 | 30 | **95** | −053 | 28 | **45** | 459 | 28 |
| **7100** | −126 | 30 | **7350** | −629 | 29 | **7600** | 081 | 29 | **7850** | −487 | 28 |
| **05** | 156 | 31 | **55** | 658 | 30 | **05** | −110 | 28 | **55** | −515 | 27 |
| **10** | −187 | 30 | **60** | −688 | 29 | **10** | 138 | 29 | **60** | 542 | 28 |
| **15** | 217 | 31 | **65** | 717 | 30 | **15** | −167 | 28 | **65** | −570 | 27 |
| **20** | −248 | 30 | **70** | −747 | 29 | **20** | 195 | 29 | **70** | 597 | 28 |
| **25** | 278 | 31 | **75** | 776 | 30 | **25** | −224 | 28 | **75** | 625 | 28 |
| **30** | −309 | 30 | **80** | −806 | 29 | **30** | 252 | 29 | **80** | −653 | 27 |
| **35** | 339 | 31 | **85** | 835 | 29 | **35** | −281 | 28 | **85** | 680 | 28 |
| **40** | −370 | 30 | **90** | 864 | 30 | **40** | 309 | 29 | **90** | −708 | 27 |
| **45** | 400 | 31 | **95** | −894 | 29 | **45** | −338 | 28 | **95** | 735 | 28 |
| **7150** | −431 | 30 | **7400** | 923 | 30 | **7650** | 366 | 29 | **7900** | −763 | 27 |
| **55** | −461 | 30 | **05** | −953 | 29 | **55** | −395 | 28 | **05** | 790 | 28 |
| **60** | 491 | 31 | **10** | −982 | 29 | **60** | −423 | 28 | **10** | −818 | 27 |
| **65** | −522 | 30 | **15** | 87011 | 29 | **65** | 451 | 29 | **15** | 845 | 28 |
| **70** | −552 | 30 | **20** | 040 | 30 | **70** | −480 | 28 | **20** | −873 | 27 |
| **75** | 582 | 30 | **25** | −070 | 29 | **75** | −508 | 28 | **25** | −900 | 27 |
| **80** | 612 | 31 | **30** | −099 | 29 | **80** | 536 | 28 | **30** | 927 | 28 |
| **85** | −643 | 30 | **35** | 128 | 29 | **85** | 564 | 29 | **35** | −955 | 27 |
| **90** | −673 | 30 | **40** | 157 | 29 | **90** | −593 | 28 | **40** | 982 | 27 |
| **95** | 703 | 30 | **45** | 186 | 30 | **95** | −621 | 28 | **45** | 90009 | 28 |
| **7200** | 733 | 30 | **7450** | −216 | 29 | **7700** | 649 | 28 | **7950** | −037 | 27 |
| **05** | 763 | 31 | **55** | −245 | 29 | **05** | 677 | 28 | **55** | 064 | 27 |
| **10** | −794 | 30 | **60** | −274 | 29 | **10** | 705 | 29 | **60** | 091 | 28 |
| **15** | −824 | 30 | **65** | −303 | 29 | **15** | −734 | 28 | **65** | −119 | 27 |
| **20** | −854 | 30 | **70** | 332 | 29 | **20** | −762 | 28 | **70** | −146 | 27 |
| **25** | −884 | 30 | **75** | 361 | 29 | **25** | −790 | 28 | **75** | 173 | 27 |
| **30** | −914 | 30 | **80** | 390 | 29 | **30** | −818 | 28 | **80** | 200 | 27 |
| **35** | −944 | 30 | **85** | 419 | 29 | **35** | 846 | 28 | **85** | 227 | 28 |
| **40** | −974 | 30 | **90** | 448 | 29 | **40** | 874 | 28 | **90** | −255 | 27 |
| **45** | 86004 | 30 | **95** | 477 | 29 | **45** | 902 | 28 | **95** | −282 | 27 |

P. P.

| | **31** | **30** | **29** | **28** | **27** |
|---|---|---|---|---|---|
| **1** | 0.6 | 0.6 | 0.6 | 0.6 | 0.5 |
| **2** | 1.2 | 1.2 | 1.2 | 1.1 | 1.1 |
| **3** | 1.9 | 1.8 | 1.7 | 1.7 | 1.6 |
| **4** | 2.5 | 2.4 | 2.3 | 2.2 | 2.2 |
| **5** | 3.1 | 3.0 | 2.9 | 2.8 | 2.7 |
| **6** | 3.7 | 3.6 | 3.5 | 3.4 | 3.2 |
| **7** | 4.3 | 4.2 | 4.1 | 3.9 | 3.8 |
| **8** | 5.0 | 4.8 | 4.6 | 4.5 | 4.3 |
| **9** | 5.6 | 5.4 | 5.2 | 5.0 | 4.9 |
| **10** | 6.2 | 6.0 | 5.8 | 5.6 | 5.4 |

## LOGARITHMS OF NUMBERS.—*Continued.*

| No. | Log. | Dif. | No. | Log. | Dif. | No. | Log. | Dif. | No. | Log. | Dif. | P. P. | |
|---|---|---|---|---|---|---|---|---|---|---|---|---|---|
| **8000** | 90309 | 27 | **8250** | 91645 | 27 | **8500** | 92942 | 25 | **8750** | 94201 | 25 | | **28** |
| **05** | 336 | 27 | **55** | −672 | 26 | **05** | 967 | 26 | **55** | −226 | 24 | **1** | 0.6 |
| **10** | 363 | 27 | **60** | 698 | 26 | **10** | −993 | 25 | **60** | 250 | 25 | **2** | 1.1 |
| **15** | 390 | 27 | **65** | 724 | 27 | **15** | 93018 | 26 | **65** | 275 | 25 | **3** | 1.7 |
| **20** | 417 | 28 | **70** | −751 | 26 | **20** | −044 | 25 | **70** | −300 | 25 | **4** | 2.2 |
| | | | | | | | | | | | | **5** | 2.8 |
| **25** | −445 | 27 | **75** | −777 | 26 | **25** | 069 | 26 | **75** | −325 | 24 | **6** | 3.4 |
| **30** | −472 | 27 | **80** | 803 | 26 | **30** | −095 | 25 | **80** | 349 | 25 | **7** | 3.9 |
| **35** | −499 | 27 | **85** | 829 | 26 | **35** | 120 | 26 | **85** | 374 | 25 | **8** | 4.5 |
| **40** | −526 | 27 | **90** | 855 | 27 | **40** | −146 | 25 | **90** | −399 | 25 | **9** | 5.0 |
| **45** | −553 | 27 | **95** | −882 | 26 | **45** | 171 | 26 | **95** | −424 | 24 | **10** | 5.6 |
| **8050** | −580 | 27 | **8300** | −908 | 26 | **8550** | −197 | 25 | **8800** | 448 | 25 | | **27** |
| **55** | −607 | 27 | **05** | −934 | 26 | **55** | 222 | 25 | **05** | −473 | 25 | **1** | 0.5 |
| **60** | −634 | 26 | **10** | 960 | 26 | **60** | 247 | 26 | **10** | −498 | 24 | **2** | 1.1 |
| **65** | 660 | 27 | **15** | 986 | 26 | **65** | −273 | 25 | **15** | 522 | 25 | **3** | 1.6 |
| **70** | 687 | 27 | **20** | 92012 | 26 | **70** | 298 | 25 | **20** | −547 | 24 | **4** | 2.2 |
| | | | | | | | | | | | | **5** | 2.7 |
| **75** | 714 | 27 | **25** | 038 | 27 | **75** | 323 | 26 | **25** | 571 | 25 | **6** | 3.2 |
| **80** | 741 | 27 | **30** | −065 | 26 | **80** | −349 | 25 | **30** | 596 | 25 | **7** | 3.8 |
| **85** | 768 | 27 | **35** | −091 | 26 | **85** | 374 | 25 | **35** | −621 | 24 | **8** | 4.3 |
| **90** | −795 | 27 | **40** | −117 | 26 | **90** | 399 | 26 | **40** | 645 | 25 | **9** | 4.9 |
| **95** | −822 | 27 | **45** | −143 | 26 | **95** | −425 | 25 | **45** | −670 | 24 | **10** | 5.4 |
| **8100** | −849 | 26 | **8350** | −169 | 26 | **8600** | −450 | 25 | **8850** | 694 | 25 | | **26** |
| **05** | 875 | 27 | **55** | −195 | 26 | **05** | 475 | 25 | **55** | −719 | 24 | **1** | 0.5 |
| **10** | 902 | 27 | **60** | −221 | 26 | **10** | 500 | 26 | **60** | 743 | 25 | **2** | 1.0 |
| **15** | −929 | 27 | **65** | −247 | 26 | **15** | −526 | 25 | **65** | −768 | 24 | **3** | 1.6 |
| **20** | −956 | 26 | **70** | −273 | 25 | **20** | −551 | 25 | **70** | 792 | 25 | **4** | 2.1 |
| | | | | | | | | | | | | **5** | 2.6 |
| **25** | 982 | 27 | **75** | 298 | 26 | **25** | −576 | 25 | **75** | −817 | 24 | **6** | 3.1 |
| **30** | 91009 | 27 | **80** | 324 | 26 | **30** | 601 | 25 | **80** | 841 | 25 | **7** | 3.6 |
| **35** | −036 | 26 | **85** | 350 | 26 | **35** | 626 | 25 | **85** | −866 | 24 | **8** | 4.2 |
| **40** | 062 | 27 | **90** | 376 | 26 | **40** | 651 | 25 | **90** | 890 | 25 | **9** | 4.7 |
| **45** | 089 | 27 | **95** | 402 | 26 | **45** | 676 | 26 | **95** | −915 | 24 | **10** | 5.2 |
| **8150** | −116 | 26 | **8400** | −428 | 26 | **8650** | −702 | 25 | **8900** | 939 | 24 | | **25** |
| **55** | 142 | 27 | **05** | −454 | 26 | **55** | −727 | 25 | **05** | 963 | 25 | **1** | 0.5 |
| **60** | 169 | 27 | **10** | −480 | 25 | **60** | −752 | 25 | **10** | −988 | 24 | **2** | 1.0 |
| **65** | −196 | 26 | **15** | 505 | 26 | **65** | −777 | 25 | **15** | 95012 | 24 | **3** | 1.5 |
| **70** | 222 | 27 | **20** | 531 | 26 | **70** | −802 | 25 | **20** | 036 | 25 | **4** | 2.0 |
| | | | | | | | | | | | | **5** | 2.5 |
| **75** | −249 | 26 | **25** | −557 | 26 | **75** | −827 | 25 | **25** | −061 | 24 | **6** | 3.0 |
| **80** | 275 | 27 | **30** | −583 | 26 | **80** | −852 | 25 | **30** | 085 | 24 | **7** | 3.5 |
| **85** | −302 | 26 | **35** | −609 | 25 | **85** | −877 | 25 | **35** | 109 | 25 | **8** | 4.0 |
| **90** | 328 | 27 | **40** | 634 | 26 | **90** | −902 | 25 | **40** | −134 | 24 | **9** | 4.5 |
| **95** | −355 | 26 | **45** | −660 | 26 | **95** | −927 | 25 | **45** | 158 | 24 | **10** | 5 0 |
| **8200** | 381 | 27 | **8450** | −686 | 25 | **8700** | −952 | 25 | **8950** | 182 | 25 | | **24** |
| **05** | −408 | 26 | **55** | 711 | 26 | **05** | −977 | 25 | **55** | −207 | 24 | **1** | 0.5 |
| **10** | 434 | 27 | **60** | 737 | 26 | **10** | 94002 | 25 | **60** | −231 | 24 | **2** | 1.0 |
| **15** | −461 | 26 | **65** | −763 | 25 | **15** | −027 | 25 | **65** | 255 | 24 | **3** | 1.4 |
| **20** | 487 | 27 | **70** | 788 | 26 | **20** | −052 | 25 | **70** | 279 | 24 | **4** | 1.9 |
| | | | | | | | | | | | | **5** | 2.4 |
| **25** | −514 | 26 | **75** | −814 | 26 | **25** | −077 | 24 | **75** | 303 | 25 | **6** | 2.9 |
| **30** | −540 | 26 | **80** | −840 | 25 | **30** | 101 | 25 | **80** | −328 | 24 | **7** | 3.4 |
| **35** | 566 | 27 | **85** | 865 | 26 | **35** | 126 | 25 | **85** | −352 | 24 | **8** | 3.8 |
| **40** | −593 | 26 | **90** | −891 | 25 | **40** | 151 | 25 | **90** | −376 | 24 | **9** | 4.3 |
| **45** | 619 | 26 | **95** | 916 | 26 | **45** | −176 | 25 | **95** | 400 | 24 | **10** | 4.8 |

## LOGARITHMS OF NUMBERS. — *Continued.*

| No. | Log. | Dif. | No. | Log. | Dif. | No. | Log. | Dif. | No. | Log. | Dif. |
|---|---|---|---|---|---|---|---|---|---|---|---|
| **9000** | 95424 | 24 | **9250** | 96614 | 24 | **9500** | 97772 | 23 | **9750** | 98900 | 23 |
| **05** | 448 | 24 | **55** | −638 | 23 | **05** | 795 | 23 | **55** | −923 | 22 |
| **10** | 472 | 25 | **60** | 661 | 24 | **10** | 818 | 23 | **60** | −945 | 22 |
| **15** | −497 | 24 | **65** | −685 | 23 | **15** | −841 | 23 | **65** | 967 | 22 |
| **20** | −521 | 24 | **70** | −708 | 23 | **20** | −864 | 22 | **70** | 989 | 23 |
| **25** | −545 | 24 | **75** | 731 | 24 | **25** | 886 | 23 | **75** | 99012 | 22 |
| **30** | −569 | 24 | **80** | −755 | 23 | **30** | 909 | 23 | **80** | −034 | 22 |
| **35** | −593 | 24 | **85** | 778 | 24 | **35** | 932 | 23 | **85** | 056 | 22 |
| **40** | −617 | 24 | **90** | −802 | 23 | **40** | −955 | 23 | **90** | 078 | 22 |
| **45** | −641 | 24 | **95** | −825 | 23 | **45** | −978 | 22 | **95** | 100 | 23 |
| **9050** | −665 | 24 | **9300** | 848 | 24 | **9550** | 98000 | 23 | **9800** | −123 | 22 |
| **55** | −689 | 24 | **05** | −872 | 23 | **55** | 023 | 23 | **05** | −145 | 22 |
| **60** | −713 | 24 | **10** | −895 | 23 | **60** | −046 | 22 | **10** | −167 | 22 |
| **65** | −737 | 24 | **15** | 918 | 24 | **65** | 068 | 23 | **15** | 189 | 22 |
| **70** | −761 | 24 | **20** | −942 | 23 | **70** | 091 | 23 | **20** | 211 | 22 |
| **75** | −785 | 24 | **25** | −965 | 23 | **75** | −114 | 23 | **25** | 233 | 22 |
| **80** | −809 | 23 | **30** | 988 | 23 | **80** | −137 | 22 | **30** | 255 | 22 |
| **85** | 832 | 24 | **35** | 97011 | 24 | **85** | 159 | 23 | **35** | 277 | 23 |
| **90** | 856 | 24 | **40** | −035 | 23 | **90** | −182 | 22 | **40** | −300 | 22 |
| **95** | 880 | 24 | **45** | −058 | 23 | **95** | 204 | 23 | **45** | −322 | 22 |
| **9100** | 904 | 24 | **9350** | 081 | 23 | **9600** | 227 | 23 | **9850** | −344 | 22 |
| **05** | −928 | 24 | **55** | 104 | 24 | **05** | −250 | 22 | **55** | −366 | 22 |
| **10** | −952 | 24 | **60** | −128 | 23 | **10** | 272 | 23 | **60** | −388 | 22 |
| **15** | −976 | 23 | **65** | −151 | 23 | **15** | −295 | 23 | **65** | −410 | 22 |
| **20** | 999 | 24 | **70** | −174 | 23 | **20** | −318 | 22 | **70** | −432 | 22 |
| **25** | 96023 | 24 | **75** | 197 | 23 | **25** | 340 | 23 | **75** | −454 | 22 |
| **30** | 047 | 24 | **80** | 220 | 23 | **30** | −363 | 22 | **80** | −476 | 22 |
| **35** | −071 | 24 | **85** | 243 | 24 | **35** | 385 | 23 | **85** | −498 | 22 |
| **40** | −095 | 23 | **90** | −267 | 23 | **40** | −408 | 22 | **90** | −520 | 22 |
| **45** | 118 | 24 | **95** | −290 | 23 | **45** | 430 | 23 | **95** | −542 | 22 |
| **9150** | 142 | 24 | **9400** | −313 | 23 | **9650** | −453 | 22 | **9900** | −564 | 21 |
| **55** | −166 | 24 | **05** | −336 | 23 | **55** | 475 | 23 | **05** | 585 | 22 |
| **60** | −190 | 23 | **10** | −359 | 23 | **60** | −498 | 22 | **10** | 607 | 22 |
| **65** | 213 | 24 | **15** | 382 | 23 | **65** | 520 | 23 | **15** | 629 | 22 |
| **70** | −237 | 24 | **20** | 405 | 23 | **70** | −543 | 22 | **20** | 651 | 22 |
| **75** | −261 | 23 | **25** | 428 | 23 | **75** | 565 | 23 | **25** | 673 | 22 |
| **80** | 284 | 24 | **30** | 451 | 23 | **80** | −588 | 22 | **30** | −695 | 22 |
| **85** | −308 | 24 | **35** | 474 | 23 | **85** | −610 | 22 | **35** | −717 | 22 |
| **90** | −332 | 23 | **40** | 497 | 23 | **90** | 632 | 23 | **40** | −739 | 21 |
| **95** | 355 | 24 | **45** | 520 | 23 | **95** | −655 | 22 | **45** | 760 | 22 |
| **9200** | −379 | 23 | **9450** | 543 | 23 | **9700** | 677 | 23 | **9950** | 782 | 22 |
| **05** | 402 | 24 | **55** | 566 | 23 | **05** | −700 | 22 | **55** | 804 | 22 |
| **10** | −426 | 24 | **60** | 589 | 23 | **10** | −722 | 22 | **60** | −826 | 22 |
| **15** | −450 | 23 | **65** | 612 | 23 | **15** | 744 | 23 | **65** | −848 | 22 |
| **20** | 473 | 24 | **70** | −635 | 23 | **20** | −767 | 22 | **70** | −870 | 21 |
| **25** | −497 | 23 | **75** | −658 | 23 | **25** | −789 | 22 | **75** | 891 | 22 |
| **30** | 520 | 24 | **80** | −681 | 23 | **30** | 811 | 23 | **80** | 913 | 22 |
| **35** | −544 | 23 | **85** | −704 | 23 | **35** | −834 | 22 | **85** | −935 | 22 |
| **40** | 567 | 24 | **90** | −727 | 22 | **40** | −856 | 22 | **90** | −957 | 22 |
| **45** | −591 | 23 | **95** | 749 | 23 | **45** | 878 | 22 | **95** | 978 | 22 |

P. P.

| | **25** | **24** | **23** | **22** | **21** |
|---|---|---|---|---|---|
| **1** | 0.5 | 0.5 | 0.5 | 0.4 | 0.4 |
| **2** | 1.0 | 1.0 | 0.9 | 0.9 | 0.8 |
| **3** | 1.5 | 1.4 | 1.4 | 1.3 | 1.3 |
| **4** | 2.0 | 1.9 | 1.8 | 1.8 | 1.7 |
| **5** | 2.5 | 2.4 | 2.3 | 2.2 | 2.1 |
| **6** | 3.0 | 2.9 | 2.8 | 2.6 | 2.5 |
| **7** | 3.5 | 3.4 | 3.2 | 3.1 | 2.9 |
| **8** | 4.0 | 3.8 | 3.7 | 3.5 | 3.4 |
| **9** | 4.5 | 4.3 | 4.1 | 4.0 | 3.8 |
| **10** | 5.0 | 4.8 | 4.6 | 4.4 | 4.2 |

# INDEX